Seven Bad Habits of Safety Management

Occupational health and safety has been a growth industry for several decades and has moved beyond the realm of the human resource department and workers' compensation claims. However, the methodologies utilised and taught within the profession have changed little since the 1930s. The industry continues to operate in a "comfort zone" and, as such, has reached an improvement plateau. This important book examines seven of these antiquated comfort zones from their conceptions to implementation and explores why they fail to achieve the desired results. This book is not a how-to but an invitation to start the conversations required for change.

Seven Bad Habits of Safety Management: Examining Systemic Failure delivers seven focused chapters outlining the comfort zones they create and their impacts on new initiatives. Each critically analyses common safety practices exploring where they came from, why they fail, and a few alternatives being discussed around the world. Case studies underpin learning that will allow the reader to revisit and revise their current programs and campaigns to realise a better return on their safety investment. The book will allow the reader to better understand the root causes of systems failures faced daily in the management of health and safety and how to confront them.

This readable and exciting text from an author with over 40 years of field experience in occupational health and safety will appeal to students, researchers, and professionals working in process safety, occupational safety, safety engineering, human resources, and business management.

Murray Ritchie is an occupational health and safety practitioner, researcher, educator, and sought after keynote speaker with 40 years of experience working with various industries, governments and NGOs. He obtained a Master of Science degree in Occupational Safety and Health from the University of Greenwich, UK, late in life following a boots-on-the-ground career in the offshore oil and gas industry. He has worked on five continents consulting to a wide variety of industries, has served on programme advisory boards for two colleges and taught as a contract instructor for the University of Alberta Faculty of Extension.

Innovation and Creativity in Health and Safety

Series Editors
Mark Jenkins, Siemens Gamesa Renewable Energy Limited, and Ben Matellini, Liverpool John Moores University

The aim of this series is to deliver books from those who think laterally rather than literally. The health and safety industry is frequently based upon the perceptions, opinions and beliefs of those who have little or no in-depth knowledge of the subject. Alternatively, their judgment of the competent is clouded by bias and cognitive dissonance to produce a solution that aligns with the expectations of others. Health and safety is not a precise science. The controls and mitigations that have been demonstrated to work in one environment fail to make an impact in another. Documents such as risk assessments and method statements are frequently copied and pasted. Employees and contractors are treated as if they were robots or computers and, once correctly programmed, they should perform the task perfectly. In the health and safety environment as well as many others the warmth, comfort and ease of determinism resonate everywhere. Just because something has always been believed, it does not make it the truth. Innovation and Creativity in Health and Safety delivers books from those who dare to think differently, those who strive to innovate, and those who believe there is a better way.

We encourage the submission of proposals to this pioneering series from authors who think that health and safety could and should be better.

Seven Bad Habits of Safety Management: Examining Systemic Failure
Murray Ritchie

Seven Bad Habits of Safety Management

Examining Systemic Failure

Murray Ritchie

CRC Press
Taylor & Francis Group
Boca Raton London New York

CRC Press is an imprint of the
Taylor & Francis Group, an **informa** business

Designed cover image: Nathan Ritchie

First edition published 2024
by CRC Press
6000 Broken Sound Parkway NW, Suite 300, Boca Raton, FL 33487–2742

and by CRC Press
4 Park Square, Milton Park, Abingdon, Oxon, OX14 4RN

CRC Press is an imprint of Taylor & Francis Group, LLC

© 2024 Murray Ritchie

ISBN: 978-1-032-52027-8 (hbk)
ISBN: 978-1-032-51838-1 (pbk)
ISBN: 978-1-003-40495-8 (ebk)

DOI: 10.1201/9781003404958

Typeset in Palatino
by Apex CoVantage, LLC

For my grandchildren Hannah, Rayah and Kyla

May we have the will to set things right

before their generation is expected to

gamble their lives at work using tools

from their grandfather's grandfather.

In memory of Alan Quilley

"Thought Provoker", Change Agent,

Educator, Writer, Consultant and Mentor

Contents

Foreword

This book is designed to be read from front to back much like a novel. It is not a reference book and offers no magic bullets in search of mythical safety utopias.

We hope that the reader will better understand the root of why health and safety management fails and, at the very least, fuels discussions, critical thinking and the willingness to remove the blind spots created by compliance and rules-based safety.

Different readers will take away different perspectives. Those working in the field can reflect on the content and hopefully walk away with a few ideas to explore and implement adapted to their specific worksites. Those working in education, curriculum development or professional qualification will hopefully find a reason to break from teaching the bad habits while students will be able to apply the reference materials to critical thinking and the confidence to question what they are being taught.

Most importantly, it is our hope that regulators will join the conversation and move beyond their comfort zones long enough to support strategic step change and risk-based safety.

Lead, follow or get out of the way.

About the Author

Murray Ritchie is an occupational health and safety practitioner, researcher, educator, and sought after keynote speaker with 40 years of experience working with various industries, governments and NGOs.

He obtained a Master of Science degree in Occupational Safety and Health from the University of Greenwich, UK, late in life following a boots-on-the-ground career in the offshore oil and gas industry. He has worked on five continents consulting to a wide variety of industries, has served on programme advisory boards for two colleges and taught as a contract instructor for the University of Alberta Faculty of Extension.

He continues to consult and facilitate joint committee alignment and zone management theory, and he is a sought-after presenter and keynote speaker presenting interactive sessions on the seven bad habits.

Murray can be reached via e-mail at murray@trilenssafety.com, and speaking inquiries can be submitted at www.trilenssafety.com

Acknowledgements

This book would not be possible without the support of Kim Moes. From her willingness to create the space and environment for me to work, often sacrificing her own space, to her ability to push me when I was willing to walk away from the project to her constructive criticisms and unwavering faith in my ability. The best life partner, roommate, frontline editor, and friend any person could wish for.

Thanks to Kirsty Hardwick, James Hobbs, and the rest of the editorial and production staff at Taylor & Francis CRC Press for your guidance, patience, and hard work behind the scenes. Many thanks to Aruna Rajendran and her team from Apex who helped in the smooth production process of the book.

Gratitude to Ciarán McAleenan, Brian Fisher-Smith, Philip McAleenan, Kim Taylor, Angela Thachuk, and Tiffany Rogowski for their shared passion for creating safe and healthy work environments.

Shout-out to Clive Lloyd, Carsten Busch, Tristan Casey, Nippin Anand, Britt Andreatta, Moni Hogg, Georgina Poole and the countless other researchers and influencers who keep the discussions and debates relevant, respectful and informative.

Last, but certainly not least, special thanks owed to Nathan Ritchie for sharing his talents in the cover design and artwork.

Bad Habit 1

Plan–Do–Check–Act: The Activity Trap

Several years ago, I was delivering a presentation at a professional development conference for a national association of safety practitioners. Part of my presentation outlined the problem with utilising the Plan–Do–Check–Act (PDCA) model as a base for safety management.

After the session, one participant's feedback stated, "Who does he think he is to question international standards?"

As a speaker and critical thinker, I always enjoy receiving negative or constructive feedback and this one sentence questioning who I thought I was is by far one of the most powerful pieces of feedback I had ever received. It confirmed what I had long believed. Most safety "professionals" have a limited understanding of the processes that have been ingrained in the safety profession by well-meaning practitioners of other disciplines.

So, as this first chapter explores the challenges of the PDCA cycle, it is appropriate that we first explore its origins and how it has played out in the world of safety management.

Part 1 History

For starters, we must all understand that PDCA is not a standard. It is simply a concept or method at best, which is often utilised as the base within many standards. We must also understand that this concept was originally designed for quality control in the design and manufacturing of products.

The cycle of PDCA is commonly referred to as The Deming Cycle, or Deming Wheel, which is also misleading and incorrect. The fact that this is taught to future safety practitioners in postsecondary education the world over is a little troubling and perhaps one of the reasons many are so stuck on it as a "standard".

To better explore the challenges many safety practitioners face when utilising PDCA, we must first look at and understand the origins and intent of the model.

DOI: 10.1201/9781003404958-1

In the Beginning

Probably, the first to show it as a circle was Walter Shewhart, an American physicist, engineer and statistician, also known as the father of statistical quality control. However, in his book, *Economic control of quality of manufactured product*[1], he originally had only three steps: first in a line and later in a circle: Specification–Production–Inspection.

In the early 1930s, a young quality engineer named Edwards Deming edited Shewhart's book and went on to become highly regarded in Japan as a teacher in the field of quality control. Later, in 1950, he renamed the steps slightly and added a fourth step making it: Design–Produce–Sell–Redesign.

Still building off the Shewhart model, he again modified the concept and presented a new version during an eight-day seminar in Japan sponsored by the Japanese Union of Scientists and Engineers. In his new version of the cycle, Deming stressed the importance of constant interaction among the four steps of Design–Production–Sales–Research.

Deming's teachings in post-war Japan were highly regarded, and many believe this early work was responsible for Japan significantly improving product quality.

Through Deming's teachings, Japanese engineers picked up the concepts developed by Shewhart and Deming.

According to Masaaki Imai, Japanese executives recast the "Deming Wheel" in 1951 and stated:

- 計画 (Keikaku) for plan; project; schedule; scheme; programme
- 実施 (Jisshi) for enforcement; implementation; putting into practice; carrying out; operation; working
- チェック, which is the English word "check" written in Japanese letters
- アクション, which is the English word "action" written in Japanese letters

Translating the Japanese version into English gives the modern Plan–Do–Check–Act.

Imai did not provide details about which executives reworked the wheel, and no one has ever claimed ownership of this revision although many credit the change to Kaoru Ishikawa, a Japanese organisational theorist, best known outside Japan for the Ishikawa or "Fishbone" cause-and-effect diagram, yet, for whatever reason, the new PDCA became known as "The Deming Cycle".

More than 30 years after Deming first revised the Shewhart cycle, he again reintroduced it during four-day seminars in the 1980s. He cautioned his audiences that the Japanese PDCA version is frequently inaccurate because the English word "check" means "to hold back".

Once again, Deming modified the Shewhart cycle, and in 1993, it became Plan–Do–Study–Act (PDSA). Deming referred to the modification as "The

Shewhart Cycle for Learning and Improvement". He described it as a flow diagram for the improvement of a product or a process, and in the 1980s and 1990s, safety was very much becoming process driven.

Over the years, Deming had strong beliefs about Plan–Do–Check–Act and clearly wanted to distinguish it from the Plan–Do–Study–Act cycle.

Addressing a U.S. Government Accounting Office roundtable, Deming was asked how the quality control cycle of PDCA and PDSA related[2]. He responded with:

> They bear no relation to each other; the Deming circle is a quality control program. It is a plan for management. Four steps: Design it, make it, sell it, then test it in service. Repeat the four steps, over and over, redesign it, make it, etc. Maybe you could say that the Deming circle is for management, and the QC circle [PDCA] is for a group of people that work on faults encountered at the local level.

In a letter to Ronald D. Moen in 1990, he commented on the manuscript for *Improving Quality Through Planned Experimentation*, co-authored by Moen, Thomas R. Nolan and Lloyd P. Provost[3] he wrote as follows: *Be sure to call it PDSA, not the corrupted PDCA.*

In response to a letter he received in 1991, he commented about a chart labelled "Plan-Do-Check-Act": *What you propose is not the Deming cycle. I don't know the source of the cycle that you propose. How the PDCA ever came into existence I know not.*

Regardless of Edward Deming's obvious frustration with PDCA, we continue to teach this concept as "the Deming Cycle". What's more, within the occupational health and safety profession we continue to view PDCA as a safety management "standard".

Deming emphasised that the steps (PDSA) should be rotated constantly, with quality of product and service as the aim. So how did we get to a place where Plan–Do–Check–Act has become a "standard" in safety?

The Challenge

We can clearly see that the origin of PDCA is rooted in product development and product quality, with the idea to constantly improve a product through a redesign so that the new product sells and keeps pace with market demands and consumer needs.

As Deming stated Plan–Do–Check–Act "is for a group of people that work on faults encountered at the local level". And let's remember that, even today, safety managers and practitioners are very fault and process driven. Perhaps this is the appeal that keeps them stuck on PDCA.

The problem is that people are not products and rules-based safety focused solely on process and desired outcomes severely limits step change and strategic improvement.

In the writing of this book, I reached out to my network of safety practitioners and asked them to share, step by step, an account of how they have utilised PDCA in their daily practice. I then chose the first 100 accounts received and applied a gap analysis to each from a risk-based approach to safety management. In all cases, steps were either missing or in the wrong order.

The most common trend was for the respondents to miss the *plan* step entirely or state a target and goal as the *plan* and rush to the *do*.

We *plan* to develop a safety management system, and we *do* policies and procedures then we *check* for compliance, and lastly, we *act* by rewarding or reacting, and then we go back and *do* more safety. There is seldom a concrete *plan*, and when there is, it is most often based on compliance.

A perfect example is in achieving the misguided goal of zero (Bad Habit 3).

The company creates a policy statement something to the effect of "No job is so important that it can't be done safely" or "It is everyone's responsibility to create an incident free workplace".

Simply starting with this well-intentioned policy statement has rushed the company to the *do*.

That aside, let's for a minute entertain the notion that the *plan* is to create an incident-free workplace. The *do* comes in when we develop policies and procedures to achieve this plan. Next, we *check* it in service for compliance (are the workers following the policies and procedures) and then we *act* by rewarding or reacting.

Figure 1.1 demonstrates the PDCA "Activity Trap" based on target zero initiatives. In this example, the only measurable improvement comes from the *act*.

Safety bonuses fill the role of both react and reward. We implement some sort of bonus scheme, and the workers are either rewarded (bonus for days incident free) or they are penalised by the reaction of losing the bonus. This, in and of itself, is very effective at achieving results, but, as we will discuss in the next chapter, it is simply "Safe by Accident"[4].

Staying with the concept of PDCA, we then utilise this faulty measurement to guide us with the next *do* and we do more of the same and simply add more policies and procedures or restate our "plan".

Many argue that this example is too simplistic and that PDCA would be better applied as the method to address specific findings arising out of the *check*. Perhaps this is closer to the truth, and certainly, this stage of the cycle better lends itself to Deming's definition of "a group of people addressing a problem at the local level".

The "activity trap" applies to pretty much anything we tackle in safety management via the PDCA cycle. It is the comfort zone that impedes strategic improvement as the *plan* has not changed in decades.

The very cyclical nature of PDCA has organisations in a state of doing, and as mentioned previously, very few safety practitioners surveyed could clearly identify the source of the Plan. So, this begs the question, "Does PDCA serve its original intent as defined by the Japanese?" and if we dig a little deeper, the answer appears to be no.

FIGURE 1.1
PDCA – The Activity Trap

Beyond PDCA

Ironically, while the safety world hitched its wagon to the wrongly titled "Deming Cycle", the Japanese continued to work on improving their quality PDCA model.

Kaoru Ishikawa, a long-time colleague of Deming, and thought by many to be the originator of the PDCA quality model, continued to explore the circle within quality management. In his book *"What Is Total Quality Control?"*[5] Ishikawa expanded the four steps into six:

- Determine goals and targets.
- Determine methods of reaching goals.
- Engage in education and training.
- Implement work.
- Check the effects of implementation.
- Take appropriate action.

Would adopting this modification into the original model change anything about the way we traditionally manage safety?

Applying it to the previous example, it looks like this.

Determining goals and targets.

- Zero Incidents

Determining methods of reaching goals and targets.

- Policies and procedures

Engage in education and training.

- Inform the workforce of the new policies and procedures

Implement work.

- Utilise the policies and procedures in practice

Check the effectiveness.

- Are the workers complying with the policy or procedure

Take appropriate action.

- React or reward

It is interesting to see the expanded model played out in terms of how we traditionally manage safety. Has anything changed in respect to a safety management model? Does this new model still apply to a "group of people working on a problem at the local level" as Deming defined?

While there is merit to the six steps versus the four, the challenge occurs when we apply them utilising the old habits of rules-based safety (Bad Habit 2) which make up our "comfort zones".

The addition of education and training is a welcomed third step; however, being stuck in our comfort zones, we spend a lot of resources training the workforce on the "what", occasionally the "how", and seldom the "why".

"Here's a new procedure follow it!"

"Here's how to use fall protection. Follow the procedure!"

What workers really need to understand is "we have this new procedure because . . ."

And this is where we come full circle back to the Achilles heel . . . the *do*.

Whether we refer to it as the plan or the targets or the goals, the plan is still far too often "target zero". As complex as safety management is, it is impossible to achieve any kind of improvement beyond statistics created "by accident" based on faulty data.

We will discuss measurement in greater detail in Chapters 3 and 4 but, regardless of the systems we measure, the reality is that if the "plan" is driven by compliance, we will continue to remain firmly planted in our comfort zones.

Rules-based safety has dictated the way we plan, measure, educate and certify since the inception of safety management as a discipline. This could well be inherited by our predecessors in human resource management who, for the most part, pioneered modern safety management. It is no wonder we see safety as a "problem at the local level" and created comfort zones around rules and compliance.

Considering the *plan* appears to be the one thing that separates workers from products you would think that the time has come to rework the model yet again.

The reality is that the continuous improvement of the PDCA model did not end with Kaoru Ishikawa. His modifications would eventually lead engineers at Toyota to lean it down and revert to a slightly modified quality cycle with the addition of a first step *Observe* meaning *Observe* the current condition.

So, what does the addition of the *Observe* mean in terms of moving us out of our safety comfort zones?

Observing the worksite, as it pertains to safety, is an important first step in moving your organisation or client out of rules-based safety comfort zones and toward strategic improvement focused on actual risk versus rules.

As a problem-solving method, observing the current condition is what is often missing in rules-based safety and the absence of it in the wrongly named Deming Cycle is most often what rushes safety practitioners and organisational leaders to the *do*. This lack of observation prior to planning often has safety teams in a constant flurry of campaign writing and poster hanging. Flavour-of-the-day based on little more than industry statistics or those developed by actuaries working for regulators focused on compensation expenses.

By observing you can begin to understand the challenges as they pertain to the operation and start to *plan* specific steps to address them individually in a strategic manner.

Don't misunderstand me. Industry statistics and injury causation models are very important but remember, PDCA, as described by Deming, is intended "for a group of people that work on faults encountered at the local level". One of the dangers of getting caught up in the "activity trap" of PDCA is that it can very quickly create moral dissidence among work groups, which in turn fuels "have to" performance (Bad Habit 5).

So, what would it look like if we start with the *Observe* in our example of "target zero"?

The reality is that nothing changes with respect to our comfort zones. In truth, "target zero" is deeply entrenched in our comfort zone and for whatever reason safety practitioners have a hard time letting it go. Simply taking the time to observe does not really change the cycle of the activity trap if we are looking for lagging indicators and basing our success on the absence of incident reports. That is to say that within the comfort zone of target zero, the observation would be either "people are getting hurt" or worse yet "no one is getting hurt" (signs you are "safety by accident").

Of course, in our use of zero incidents as a plan observing the operational reality prior to planning only works when we drop the plan, which in this case is almost always to develop more policy.

What's missing is something to move us out of our comfort zones and get us focused on a strategic *plan*.

Alternative Approaches

One relatively new model worth introducing here is the "Praxis Model"[6], which lends itself well to eliminating the challenges of simply starting at a *plan* that rushes us to the *do*.

In this model, developed by Ciarán and Philip McAleenan in 2015, the first step is to choose a work task or, perhaps in the case of target zero, a social activity. Make no mistake about it, if you have drank the "goal zero" cool aid you are about to embark on a journey of sociology.

So, for argument's sake, let's say that "zero" is your social activity, or lack thereof. The next step then is to observe and analyse. This now becomes dependent on what it is you observe. In most cases, the safety practitioner will observe incident rates. Why? Because that's the comfort zone they have been trained to observe. This is why the next step becomes so very important.

Critical reflection, when applied from a risk-based vantage point, will almost always point you to the realisation that, regardless of the numbers, you have in fact not hit the target.

My favourite example of this is the organisations who tell me they have reached zero harm targets, including first aid cases yet, when you visit the job site you find first aid kits that are depleted.

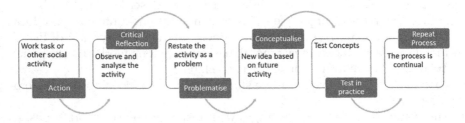

FIGURE 1.2
Praxis: Defining Model Managing Safety, Health, Wellbeing

Sticking with this example, you move to the next step which is to restate the social activity as a problem. In this case, the problem is not target zero. The problem is people are getting hurt and not reporting.

Instantly here, we have moved away from "zero" and identified a problem that we can address and measure.

The next step then takes the conceptualised findings and applies them to future activity. It is the wording of "future activities" that should, if implemented correctly, start to move us out of our activity trap comfort zone.

In the chapters ahead, we will explore pro-active planning and the use of leading indicators but, sticking with this model and the target zero analogy, once we have shifted our language of zero and framed it as a problem, we are able to develop concepts and plans to eliminate the described problem. Again, in this case the problem is that workers are being injured and not reporting. The most common side effect of rewarding targets zero.

The challenge of course is that often we remain in our comfort zones, and in this example, our "new idea" or concept will often come in the form of zero tolerance for not reporting or the development of "life saving rules", which quite likely created the problem in the first place.

This model addresses the challenge of comfort zones in the last step of "test in practice". That is to say that we are not simply checking for compliance we are testing our concept in service at the job site not against desired outcomes. It has either addressed the problem or it has not.

This, in theory, takes our focus off the worker and places it on the problem. The problem is that workers are not reporting because we reward them not to or some other form of consequence that drive the reporting underground.

While the model is not cyclical in nature, it is continual. As a continual process within our example, the concept has either worked to address the problem, in which case you are now receiving reports with which you can then form the basis of your next task or social activity to assess, or, it has not addressed the problem in which case we re-examine the task, or in this case social activity, and do a deeper dive into analysing the situation and redefine the root cause of the problem.

Keeping with our example of people getting hurt and not reporting, further analysis may lead us to framing the problem as "people don't report because the system is cumbersome" or perhaps, "people don't report because of the negative consequences that come from reporting" and so on. In practice, this better fits Deming's description of "a group of people that work on faults encountered at the local level".

This now identifies another problem whereby a plan can be conceptualised and put into action. Again, if it works problem solved. If it does not, dig deeper into framing the problem and try something new.

Restating the problem as required ensures we move our focus out of our comfort zone and avoid the activity trap of Plan–Do–Check–Act.

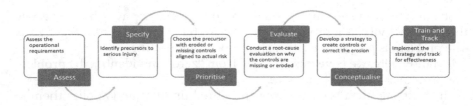

FIGURE 1.3
ASPECT: Model for Strategic Step-Change

Another similar model that came about during the time that McAleenan and McAleenan were creating the Praxis model was Assess–Specify–Prioritise–Evaluate–Conceptualise–Train/Track (ASPECT)[7].

Originally designed by Tri-Lens Safety as a method to improve operational environments, the model focuses on defining and prioritising challenges one at a time based on operational realities and leading indicators as opposed to impossible targets such as zero. That is to say that the model turns the organisation's attention away from an invisible destination and focuses on the journey.

The ASPECT method starts with a real-time assessment of the organisation's operations. A safety case of sorts whereby the everyday work requirements are assessed based on real-time hazards/risks/controls.

The second step is to specify which of the work requirements have the potential to cause serious injury or loss if not properly controlled. Not unlike a hazard identification/risk assessment. The intent here is to draw our attention away from the comfort zones of personal protection equipment (PPE), slips trips and falls, pinch points and other flavour-of-the-day and focus on the operational realities.

Step 3 requires the organisation to examine the list of serious injury precursors and identify which of them are either missing controls (e.g. confined space entry process not developed) and/or which of the already developed controls have been eroded at the worksite (e.g. fall protection procedures not consistently utilised and enforced). From this list, the organisation can prioritise specific improvement opportunities based on risk rather than rules and/or lagging indicators (Bad Habit 3).

Once the target has been identified, the next step is to evaluate the challenge. The challenge could be that there has never been a control system in place, perhaps the hazard is new or has not been identified until now or there is a disconnect in the training for the task or the original control is not fit for purpose or has been eroded through a lack of consistent leadership, etc.

Step 5 requires the organisation to develop a plan to address the challenge. This includes realistic corrections of erosions (if they have been identified), developing a new control or directive (based on risk, not rules), any management of change process that needs to be applied to ensure the new plan is not competing with the old policies, etc.

The last stage of the model is to train everyone in the organisation on the system or directive, ideally from the top down, and gain input and buy-in at all levels. Once the plan has been communicated to everyone involved, it then must be tracked for effectiveness.

This is not a circular model. There is a definite start and stop point. The stop point must be clearly understood and adhered to, and results must be agreed upon before moving on to the next challenge.

Time consuming perhaps but nowhere near as time-consuming as the activity trap of comfort zones and continuing to do the same things expecting different results as is often the case with PDCA.

So how does the ASPECT model fit within the Plan–Do–Check–Act model? It was specifically designed with creating a strategic step change in safety performance and, like the Praxis model, ASPECT requires measurable information long before the "plan" is considered. The focus in both these systems is on evaluation and measurement.

The last method I want to share is the Collaborate–Assess–Respond–Evaluate (CARE) model. CARE was created as a simplified version of ASPECT and came about during a special project I was involved in recently while consulting to a not-for-profit organisation.

Within the scope of work, I was tasked with creating a "safety management toolkit", which could be utilised by the communities and businesses supported by the not-for-profit organisation. My mandate was to go beyond the legal requirements and create something that supported consultation in the development of the systems.

With the ASPECT model in mind, I wanted to incorporate the intent of the Internal Responsibility System (IRS) and still ensure a top-down approach so as to not create yet another "blame the worker" management system.

I began exploring ways to rework the ASPECT model to better represent the spirit of Joint Health and Safety Committees where everyone has a voice in the identification, assessment, and control of hazards. Where everyone shares responsibility, and hazards are controlled based on risk, not rules. My goal was to create a resource that the end users could identify with and thus create a sense of personal ownership among the users.

I began by exploring the challenges of blind spots in traditional safety management and how, by design, we did safety to people as opposed to letting people do what they know to be safe. I was specifically drawn to how the activity trap of PDCA creates barriers to implementing the intent of the IRS which, requires every worker to be responsible for their own safety and the safety of others. It is the comfort zones, created in part by the PDCA cycle, that trap the safety practitioners and human resource managers in the *do* which leads us to *do* safety to workers rather than allowing them to work safe.

Around this same time, I was presenting at a professional development conference and in conversation with like-minded practitioners I asked, "If we all were busy studying and identifying these roadblocks and blind spots and

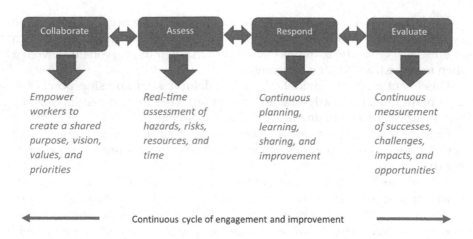

FIGURE 1.4
CARE Model

being asked to share our findings in the professional community, why has nothing changed?" The unanimous answer among the group was "Change is uncomfortable".

It was from this realisation that the system of CARE was born.

The acronym CARE represents four steps, which, unlike PDCA, do not work independently but rather collaboratively in a continuous back-and-forth cycle of engagement and improvement.

Unlike previous models, the CARE system is meant to engage everyone involved or impacted by an organisation's operations. No one element is stand alone and, as Edward Deming instructed in his PDSA model in the 1950s, all four elements require constant interaction with the others. That is to say that the impact one element has on the others should be constantly evaluated at every step.

Upon development of the CARE model, we embarked on a trial implementation within a government initiative designed to bring life to paper safety management systems. The intent of the initiative is to get safety systems out of the binders and into the hands of those who use them and/or identify the blind spots or roadblocks keeping them on the shelf. The initiative required implementing five elements a year to ensure that the challenges identified were corrected before moving on to the next five.

In the spirit of CARE, we gathered all the stakeholders and began identifying improvement opportunities within the base elements and then designed or redesigned the systems' content to meet and exceed the regulator's requirements. Once the systems were developed, we began live training for all staff where the concepts were shared and input from all levels gathered.

Each element had about a two-month cycle before the next element was addressed. At all stages of the implementation, workers were able to reflect and report on how the previous element was working for them and adjustments were made to integrate the previous module(s) with the next.

Utilising the continuous cycle of engagement and improvement while moving between the Collaborate–Assess–Respond–Evaluate, we were able to bring each system into practice and create a sense of ownership within the organisation.

The final step of the one-year cycle was to report the findings and progress, along with an action plan for improvement, to the regulator. In this first year, the first five policies were developed or amended to meet or exceed the programme requirements while the associated procedures were developed and designed in collaboration with the workforce.

One of the major blind spots we were able to identify was a disconnect between system development and board approval. As a not-for-profit organisation, all policies had to be approved at the board level and the board of directors' meetings were held quarterly, which created a gap in implementation and improvement.

Realising this the completed cycle included an action plan to have the board of directors amend their terms of reference and governance to allow senior management to approve policies and procedures related to occupational health and safety. In this way, the early successes of the programme could be built upon and carried forward to the next 12-month cycle. In short, the CARE model served to unearth blind spots that created challenges that had been slowing the improvement process and keeping the systems on the shelf.

This step-change process, fuelled by the regulator's proactive initiative, forced us to take our eyes off compliance long enough to identify a systems error not directly related to safety management but one that was having an unintended impact on all management systems. It provided an improvement opportunity that would impact all proactive measures going forward in the programme.

Conclusion

Despite the work that came before and after the development of PDCA, the safety profession's approach appears to be stuck in the 1930s and 1950s.

If the safety profession has a chance of advancing beyond the many improvement plateaus chasing zero, then safety practitioners, human resource managers, educators, regulators, supervisors, and managers will have to abandon their comfort zones.

In the remaining six chapters, we will explore bad habits that relate to PDCA and in combination create blind spots and comfort zones in a never-ending vicious circle of busy work for safety practitioners.

We will circle back to the "activity trap" often as it represents the root of why we so often fail to correctly implement new ideas and concepts. Exposing the blind spot of the constant need to "do" something will allow us to re-examine old concepts in practice as opposed to theory and implement new concepts beyond checklists and faulty measurements.

Bad Habit 2

Safety Management Systems: Rules vs. Risk

In Chapter 1, we looked at the systemic failure of the PDCA Activity Trap. In this chapter, we will expand on this to unearth the impact it can have creating systemic failures in safety management and specifically those we try to manage.

While the science of accident prevention has changed significantly over time, the profession's approach to implementing new concepts has not. How is it that a profession which demands continual improvement can be so stuck in methodology?

Carsten Busch[8] describes it best when he writes:

> The Safety Profession has a problem. Over the past decades, it has been a fertile feeding ground for Myths, Misconceptions and Misunderstandings.
>
> Pyramids, dominos, ratios, certification, zeros, absolutes, rules, audits, positive mind-set, culture change, observation schemes, checklists, best practices, slogans, Safety First, errors, root causes and risk matrices. You name it and someone will probably have twisted it into something that it should not be, through mechanisms like: visions-turned-goals, tools out of context, black and white thinking, means-becoming-the-goal, rituals without proper understanding, correlation instead of causation or belief in Silver Bullets.

Despite hard work at local, national and international levels to standardise education, certification and accident prevention methods, the profession itself is very disjointed and in practice stuck in the 1930s. We have passed through four distinct eras in safety management. The technical era, the human era, the organisation era and the systems era and we are now in what many refer to as the holistic era and, while the ideologies have changed in theory, the implementation remains somewhat the same in practice.

Imagine NASA developing the means to achieve return flight into space and Elon Musk or Richard Branson attempting to replicate it using a single-engine aircraft. While the principles of flight have not changed, and many of them apply to space travel, you can neither rely solely on the old ways to achieve new heights and likewise you cannot discard them. In occupational health and safety that is exactly what we do and have done for over a century. This is not to say that we should throw the baby out with the bath water but that we need to learn the difference.

DOI: 10.1201/9781003404958-2

During the industrial revolution, and leading up to the 1900s, those managing safety were focused on the new technologies being utilised in the workplace. Machinery was viewed as the major contributor to accident causation as it was believed that most accidents occurred when technology failed, which led to the mindset that if you could keep the machines safe, the workplace would be safe.

This approach to safety management became known as the *"Technical Era"* and led to marked improvements such as guarding, systems redundancy, maintenance and other methods of engineering hazards out of the workplace.

As the 1800s came to an end, regulators and industrialists, realising limited success in accident prevention and rising costs of incidents, turned their attention to the worker. The belief then became that if the machinery is safe the worker must be the problem, and the adage of safe machine = safe plant, became safe worker = safe workplace. A sentiment that is alive and well today.

Leading up to this assumption, in the 1800s, there was a shared belief that alcohol contributed to worker incidents. This narrative was fuelled by the growing temperance movement in the UK[9] and would eventually spill into the US leading to the first temperance law in Massachusetts in 1838 and state-wide prohibition laws in Maine in 1846[10].

By the turn of the century, citing limited success policing alcohol, the authorities and industrialists enlisted the help of psychologists to study the human at work and identify the human impact on accident causation.

Early psychological research was confirming that all accidents could be prevented if only the worker could be trained to do so. This, in turn, supported the insurance companies' investigators who were publishing findings that indicated statistics such as 90% of root causes in workplace injuries were worker owned and the remaining 10% or less attributed to unpreventable conditions or acts of god.

These early findings expanded to the popular belief that some people are simply accident prone and, by the 1930s, it was generally agreed that the worker was the problem and should be managed in the same manner as technology.

In a 1931 study, Herbert W. Heinrich, an Assistant Superintendent of the Engineering and Inspection Division at the Travelers Insurance Company in the United States, reported the root causes of 1490 lost time and/or medical treatment incidents. The numbers were reported as:

- 975   accidental human error
- 312   carelessness (255 by injured person and 57 by co-worker)
- 118   complications from unreported wounds
- 38   miscellaneous (root cause undetermined)
- 34   failure to wear safety glasses
- 13   defective or unguarded equipment

In total, 1,477 were considered worker error while only 13, less than 1 percent, were considered employer related. With a continued focus on the worker, Heinrich suggested what he called the "fundamental principles for accident prevention". These were as follows:

1. The creation and maintenance of active interest in safety
2. Fact Finding
3. Corrective action based on the facts.

There are many early concepts attributed to Heinrich, as we will explore in the chapters ahead, but oddly, this is the one that we somehow abandoned yet, in practice, this may be the only one we should have adopted. Part of the reason for this may have resulted from the mindset of the day when observing the fundamentals in practice. Having tested these principles, Heinrich reported[11]:

> *Management may have entire confidence that the accident-prevention-methods herein advocated are practical and may be applied successfully. As a matter of fact, they are not based on theory as much as on time-proved practice. They represent an orderly and logical sequence of steps that have been taken successfully in the past and are now being taken wherever satisfactory results are more than fortuitous.*
>
> *In many instances accidents prevention methods are adopted without conscious selection, sound reasoning, or knowledge that they fit the case to which they are applied – and very often they work satisfactorily.*

His in-field observations are interesting in that he is in part attributing common sense to the success. To understand his interpretation, it may be important to define what is meant by "common sense". Simply stated, common sense is common experience, which, in the early 1930s, was very relevant. At a global level, many managers and supervisors had a shared experience fighting in the first world war and understood discipline and rules. At local levels, the norm was for people to stay in the town they grew up in and take over the family farms and businesses or work in the family trade or profession. This would account for a large body of knowledge that was taught and shared and therefore common among them.

Today, this could also be the case within a specific trade or profession. Bricklayers have common experiences that structural engineers do not share so the common sense of the bricklayer works, until you put a structural engineer into the picture, and vice versa.

In support of fundamental principle number one, the creation and maintenance of active interest in safety, Heinrich cited a case study he conducted in a large manufacturing plant where he observed:

> *In one case of record the general superintendent of a large construction company issued and followed up a single executive order and thereby reduced his accident frequency and cost over 30 percent.*

Becoming alarmed because of the rising trend of accidents experience, he called
a meeting of all superintendents, their assistants, and the foreman and stated in
no uncertain terms that he 'wanted them to stop accidents'.

This "order" would be repeated several times and just like that, in 1936, Herbert W. Heinrich unknowingly described being Safe by Accident[4].

In practice, this superintendent's actions aligned with the three principles of accident prevention. The creation and maintenance of active interest came in the form of the superintendent telling the workers, "in no uncertain terms", that accidents will stop, and he repeated this clear expectation several times. Fact finding, in this case, was driven by the reality that accidents were on the rise and costing the organisation time and money. The corrective actions were his top-down directive, which would have included the "no uncertain terms" practised in the form of discipline for those not complying and perhaps bonuses for those who did.

While this executive vice president's approach is now known to be ineffective the reality in the 1930s was that it was a common tool left over from a common experience with military practices. While the well-meaning actions of the superintendent are flawed, the sad reality is they still form part of many safety practitioners' toolkits today.

This bad habit was the result of taking a theory based on scientific findings, in this case the three fundamentals, and applying them to the comfort zone of the way we have always done things. You might say that, in the 1930s, discipline for noncompliance was "common sense".

At this point in time, the safety profession would have been well served by keeping with the three fundamentals and simply monitoring the corrective actions to the realisation that discipline impedes fact finding and evolving the methods from there staying in line with the original three principles. Had these become the comfort zone in the 1930s perhaps new concepts arising in safety research today would have a better chance of being implemented in context rather than dismissed and allocated to the myth, misconception and misunderstanding file.

Until this point, worker safety was dominated by the engineering discipline. This would account for Heinrich's black-and-white approach and corrective actions based on fact as this would have been common sense in the engineering discipline. The same experiences were applied in the technical age of machine management. So where did this simple concept become complicated and twisted into something that it should not be?

The answer to this lies in what could be the root inadvertently planted by Heinrich in his study of accident causation. In Chapter 1 of *Industrial Accident Prevention: A Scientific Approach*[11], he describes a meeting in the early 1930s between a manufacturing organisation's executive vice president, their treasurer and their insurance manager, concerning a review of their plant's incident costs and potential accident causations. Having heard enough bad news regarding the company's rising accident costs the vice president adjourned the meeting, just 15 minutes in, directing his team to fix the problem. In this regard, he reportedly said as follows: *As far as I can see, we must use exactly the same methods that we already employ to correct unsatisfactory conditions regarding quality and volume of product.*

It could be that this statement, captured and endorsed by Heinrich, is what contributes to the habit of taking tools out of context from outside professions and making them our own. Keep in mind that at this same time manufacturers were adopting the Japanese quality models derived from the work of Deming and Shewhart. This may well be the beginning of PDCA taken out of context.

Modern safety management has been described as: *A multidisciplinary area which draws heavily from fields as diverse as economics, engineering, industrial relations, law, management, occupational hygiene, occupational medicine, psychology, and sociology*[12].

While there is value in sharing lessons and concepts from other disciplines, we run the risk of implementing them out of context, such as PDCA, inherited from quality control, rules-based management or Taylorism[13], inherited from human resources, or concepts such as "risk = probability × outcome" borrowed from economics. While these concepts have merit, they do not align well with the original three fundamentals.

It is the comfort zone of methods taken out of context from these other disciplines that contribute to the systemic failure of Bad Habit 2, management systems based on rules and out-of-context risk management tools.

To peel back the layers on accident prevention, we will divide the elements and associated practices that came together to create Bad Habit 2 into three categories:

1. Safety Management Systems
2. Rules and
3. Risk

It is the combination of these three elements that has fuelled the many myths, misconceptions and misunderstandings. Well-intentioned executives continue to search for the magic bullet while safety managers continue to dictate accident prevention from afar and researchers and paper theorists continue to rework old bad habits into new ones.

Safety Management Systems

In this section, we will look briefly at safety management systems (SMS) and how they relate or depart from Heinrich's three fundamentals of *creating and maintaining active interests in safety, fact finding, and corrective actions based on the facts.*

The Health and Safety Executive (HSE) in the UK updated their publication "Successful Health and Safety Management"[14] in 2006. When describing the changes, they wrote:

> *The revision does not alter the basic framework for managing health and safety set out in earlier editions, which received widespread acceptance.*

The "basic framework" referenced in the publication "Successful Health and Safety Management" first published in 1991 outlined the key elements of successful safety management as:

1. Policy
2. Organisation
3. Planning
4. Measuring performance
5. Auditing and rereviewing performance

This seems to be somewhat of a departure to the three fundamentals outlined by Heinrich, but could his fundamentals fit into the five elements of the basic framework? Or more importantly, was there any consideration to do so?

To answer this, it is important to understand where the basic framework came from and why.

As early as 1910 Fredrick Taylor, an American engineer, began exploring productivity in manufacturing focused on efficiency in the hopes of doing away with *"natural laziness"*. We will dive deeper into "Taylorism" as we explore rules-based safety in the next section, but worth mentioning here is that, while Taylor was not interested in workplace safety, he did recognise connections between management responsibilities and workplace accidents. These connections were all but lost as the technical age of safety was transitioning into the *"Human Era"* and most safety researchers were content with blame the worker causation theories.

In the 1960s, governments began passing endless amounts of legislation in North America, the UK and various European countries. In the United States, most of these laws were specific to certain industries, such as the Federal Coal Mine and Safety Act or the Contract Work Hours and Safety Standards Act, etc. This would lead to the development of a Federal Occupational Safety and Health Act (OSHA) in the United States in 1970, which was seen as the most significant milestone in the history of American safety regulations to date[15].

Motivated by major events in the 1980s, such as the Chernobyl nuclear disaster in Ukraine, the Exxon Valdez tanker spill off the coast of Alaska, and the Piper Alpha offshore platform explosion in the North Sea, investigators began promoting the idea that it was not enough to simply focus on technology and human error and the belief now became "if the organisation is safe, then we will be safe"[16]. This marked the beginning of the *"Organisational Era"*.

Total Quality Management (TQM)[17], a method developed to enhance quality and productivity in business organisations, was introduced to the safety profession in the form of Total Safety Management (TSM)[18] in 1998.

The term Total Quality Management was first coined by the US Naval Air Command. This method would form the original elements of the ISO 9000 series for quality management based on the work of Walter Shewhart and Edward Deming's methods for statistical analysis and control of quality from the 1930s, which would become PDCA.

TQM focused on the organisation and, as a performance-oriented approach, was considered the next magic bullet for safety in the form of TSM. This led to seeing safety as a component of management systems that could incorporate quality, health, environment and later security.

This belief, coupled with yet more major incidents, such as the 2003 Space Shuttle Columbia re-entry disaster, led regulators and causation researchers to the realisation that developing a safe organisation required more than isolating causes at the organisational level or managing humans and engineering technology. Armed with the knowledge that it was the complexity and interdependence of technology, humans and organisations that played a critical role in accident causation, the safety profession entered its fourth stage, *"The Systems Era"*.

The International Organization for Standardization, in 1996, developed ISO 14001 standard for environmental management. At the same time, the British Standards Institute (BSI) had released BSI 8800 "Guidelines for Health and Safety Management Systems"[19], which would provide guidelines for the development of the Occupational Health and Safety Assessment Series (OHSAS) 18001 released in 1999.

OHSAS 18000 series, while intended to be an international standard, became a starting point for those nations who wished to utilise their own standards. Countries like Canada (CSA Z-100) and the United States (ASA Z-10) amended the BSI standard not by dismissing it but by adding elements to it in an attempt to make it their own. Others, like Australia and several European, Asian and African countries adopted OHSAS 18001.

In 2007 OHSAS 18001 was updated to better align with ISO 14001 (environment) and ISO 9001 (quality) and was adopted by the UK as their national standard (BS OHSAS 18001).

On March 31, 2021, the BS OHSAS 18001 standard was replaced with the new ISO 45001 standard for occupational health and safety management systems.

The new standard was developed in consultation with more than 70 countries with the OHSAS 18001 as the base and with the British Standards Institution (BSI) serving as the committee secretariat[20]. The goal was to create a standard that would allow for easy transfer of certification from various national standards and alignment to other ISO standards (quality, environment, etc.) with an eye on streamlining the auditing process.

The influence OHSAS 18001 had on the new ISO standard and, prior to that, many national standards around the world, cannot be understated. Many elements of the Health and Safety Executive guidelines, as referenced above, found their way into ISO 45001 as did those from previous ISO standards like those for quality and environment.

While the standard, unlike many that came before, does not specifically outline the use of PDCA, those charged with implementing it fell back on this comfort zone.

This mindset is most likely a hangover from the Health and Safety Executive whose guidelines for "Successful Health and Safety Management" were that PDCA is the way we've always done it. In their guidelines, first published in 1991 the Health and Safety Executive demonstrate their five elements of safety management as they align to PDCA[21]. (see Figure 2.1)

In an online guide to OHSAS 18001 written in 2017 by Bertrand Duteil[22], he described the BSI standard in relation to PDCA as follows:

> *The OHSAS 18001 is based on the principle of steady improvement and plans the introduction of work safety and health protection management system. The principle of steady improvement is based on the PDCA cycle, also known as the Deming wheel or the Shewhart cycle.*

The writer went on to describe how PDCA fits in the BSI standard by writing:

Plan: *Safety policy and planning –* **Do**: *Realisation and operation*

Check: *Review and certification –* **Act**: *Management evaluation*

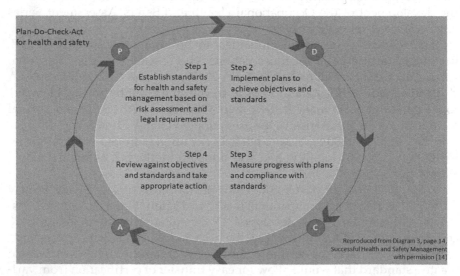

Plan-Do-Check-Act for health and safety

Step 1
Establish standards for health and safety management based on risk assessment and legal requirements

Step 2
Implement plans to achieve objectives and standards

Step 4
Review against objectives and standards and take appropriate action

Step 3
Measure progress with plans and compliance with standards

Reproduced from Diagram 3, page 14, Successful Health and Safety Management with permision [14]

FIGURE 2.1
PDCA for Safety Management as promoted by the Health and Safety Executive in 2009

The need to hold fast to the PDCA comfort zone has, in many cases, skewed the intent of the ISO standard. A quick Google search can demonstrate how the original intent, to create a standardised approach to organising a safety management system, has turned into an attempt to turn the standard into the management system.

The intent defined by the International Organisation for Standards[23] states as follows: *ISO 45001:2018 specifies requirements for an occupational health and safety (OH&S) management system and gives guidance for its use.*

In other words, the standard provides a template outlining the types of things that must be included in an SMS such as previsions for continual improvement, fulfilment of legal requirements and other requirements, or a means of achievement of objectives. What it does not do is dictate what those must be.

In practice, many well-meaning safety practitioners have used the standard to create a one-size-fits-all safety management system and when caught up in the activity trap find themselves interpreting intent as content.

In a recent article written by Jona Tarlengco[24], published online by SafetyCulture (www.safetyculture.com) in December of 2022, she describes Safety Management Systems this way:

> *Safety Management System (SMS) is a collection of structured, company-wide processes that provide effective risk-based decision-making for daily business functions. . . . SMS can also serve as a formal means of meeting statutory requirements . . . the key processes of a safety management system are hazard identification, occurrence reporting, risk management, performance measurement, and quality assurance.*

What the new ISO standard offers users is a means to organise this collection in a way that the system can be easily audited and certified. What it does not do is guarantee regulatory compliance, hazard identification, risk management or a quality (fit for purpose) system.

The ISO 45001 standard is a great accomplishment in the evolution of safety management, yet because so many continue to utilise tools out of context in implementation and misunderstand the intent, it all too often becomes a paper chase in search of compliance and certification based on rules and risk management techniques beyond the average workers' control.

This mindset carried over to ISO 45000 despite there being no specific mention of PDCA as a requirement in the new standard. In his online Blog *"ISO 45001 Requirements and Structure"*[25], John Nolan writes as follows:

> *The biggest challenge is ensuring that the procedures, policies, and activities that are undertaken on the OH&S management system complement each other and that your system structure is correct, effective and able to be improved. This can be achieved by using the "plan, do, check, act" cycle that remains central to the standard.*

The focus on policy and procedure complementing each other often distracts from the need to ensure they are effective and continuously improving. When PDCA is added to the mix, it creates blind spots causing organisations to focus more on compliance resulting in rules-based management systems.

Rules

Regardless of the standard being utilised, or the method employed to develop it, the one remaining constant is the goal of compliance and to be compliant with anything there must be measurable targets. In health and safety management, these have always been, and continue to be, driven by rules.

In the Institution of Occupational Safety and Health (IOSH) publication Principles of Health and Safety at Work[26], Allan St John Holt writes:

> *Safety management is also concerned with influencing human behaviour, and with limiting the opportunities for mistakes to be made which could result in harm or loss. To do this safety management must take into account the ways in which people fail (ie fail to do what is expected of them and/or what is safe)*

In other words, the system must be rules based to influence behaviour and limit mistakes and include corrective actions when people fail to do what is expected.

The obsession with policing away problems is not unique to the safety profession, but no one embraces this approach more than regulators, training centres and organisations looking for a quick and easy due diligence defence for their toolbox. This is also not new and is another comfort zone that we have clung to since the technical era.

In the late 1800s, it was generally agreed that if managing machinery was not enough to stop workers from being hurt then the problem was the worker, and the fledgling safety community turned their attention toward the management of people, which brought the human resource profession front and centre to the safety table to join the engineers.

Those tasked with turning machine safeguarding into worker management needed a starting point, and with an eye focused on safeguarding machinery, it was generally agreed that organisations simply had to support the engineered controls with administrative controls on workers. This would lead to the initial philosophy of the Three E's in safety:

- Engineering
- Education and
- Enforcement.

If discipline and rules-based training were seen as the required missing piece to creating safe work, then who better to turn to for guidance than the military? Many of the early researchers and managers had returned home from the "Great War", and regardless of what side they were on, they shared an understanding of the chain of command system. A process built on the belief that things go wrong when people fail to follow orders.

This was supported by a growing belief in the insurance industry that:

1. Preventing accidents could create savings in insurance, medical treatments and lost productivity.
2. Improving engineering could prevent accidents.
3. Safety rules can be established and enforced and
4. Workers are willing to learn and adhere to safety rules.

Industrialists supported this belief because it meant that blaming the worker could save money while taking the focus off management and unsafe work conditions. Legislators and regulators supported number two as it confirmed their comfort zone established in the technical age and the fledgling human resource profession gravitated to numbers three and four as it gave them a starting point to tackle this new responsibility called safety management.

In the 1930s, human resource professionals drew a great of deal their practice from the work of Frederick Taylor, the father of scientific management which would become known as "Taylorism".

Taylor, an American engineer, first published his concepts in 1911 in the publication entitled Principles of Scientific Management[27], which would unwillingly set the tone for safety management over the next 100 years.

The Taylor control model required setting up elaborate organisations and systems and constituted one of the first formal divisions between those who do the work (workers) and those who supervise and plan it (managers).

His theory revolved around the belief that rules were most effective when work was limited to tightly specified job descriptions, limiting training to specific job functions, and limiting communication to a need-to-know basis. This created a dramatic loss in skill level and autonomy and an increase in supervision and scrutiny.

His main reason for developing scientific management was that he wished to do away with "soldiership" or "natural laziness", as he believed that all workers spent little of their time putting in full efforts. He recognised problems with worker cooperation, gaining worker consent or buy-in, perception of instruction or capacity, and the challenge of creating shared understandings and beliefs among teams. In addressing these challenges, he promoted the development of systems to provide workers with special incentives to obtain their best efforts. The concept of utilising incentives to get desired results would appeal to early safety managers and continues to plague the safety profession to this day. We will examine this comfort zone in more detail in Bad Habit 3.

It could be said that Taylor's unintended contribution to occupational health and safety was his system whereby:

> *every labourer's work was planned out well in advance, and the workmen were moved from place to place by the clerks with elaborate diagrams or maps of the yard before them, very much as chessmen are moved on a chessboard*

The clerks described here were intended to increase productivity and may well have been the genesis of the modern-day project manager. However, when applying this role to safety utilising the same toolbox, we have the safety manager writing or collecting (copy-cut-paste) policy, procedures, job hazard analysis and the like (the elaborate safety maps).

Strongly influenced by the work of Frederick Taylor, German theorist Max Weber developed the Bureaucratic Management Theory[28] in 1921. While Taylor focused on individual efficiency, Weber explored organisational structure and work environment. With a focus on individual workers, in relation to organisational structure, he proposed dividing organisations into clear hierarchies supported by detailed rules and a strong chain of command. He also promoted the development of detailed operating procedures for routine tasks.

Combined, these organisational structures and rules would serve to control the workers. This also created the need to develop a specialised workforce of trained administrative personnel who could link worker capacity and activity to productivity.

These specialised administrators may have represented early human resource managers, but, when applied to safety management, the role quickly became that of the not so specialised "safety officer".

The application of Taylorism and Bureaucratic Management Theory by safety practitioners is one of the first, and best, examples of "tools out of context".

In the 1930s Taylorism, alongside Weber's bureaucratic management theory, would influence not only those practising human resource management but also researchers and practitioners from other disciplines whose work would eventually find its way into safety management. Edwards Deming reportedly said that Taylor's principles were the foundations of his own management theories. You will recall from Bad Habit 1 that the quality control concept PDCA was based on the early teachings of Deming.

It may well be a combination of PDCA influenced by Taylorism and Bureaucratic Management Theory, all taken out of context, that creates the activity trap we witness when applied to safety management. As shown in Figure 2.2, the "plan" becomes developing elaborate organisations and systems. The "do" gets consumed by developing detailed safety rules and operating procedures. Then, we deploy specialised administrators in the form of safety officers to "check" for compliance, and we finally have our clerks provide incentives and/or plan more elaborate systems.

The initial rules-based safety approach continues to dominate the legislation and, in turn, many postsecondary education curriculums and skill set requirements for professional designations. Regardless of research and advancement made within the profession, we continue to apply them out of context being stuck in the comfort zones of the 1930s ideologies based on late 1800s teachings.

Taylor believed that it was the role and responsibility of manufacturing plant managers to determine the best way for the worker to do a job and to provide the proper tools and training. This practice still dominates safety management in the form of policies and procedures being written by safety managers detached from the worksite (Taylor's clerks) policed by onsite safety officers (Weber's administrators).

My career in the oil and gas industry began during the first Gulf War when I joined Safety Boss as a medic to support the Kuwait fires project and crossed trained as an hydrogen-sulphide (H_2S) technician. I was impressed how the team would regroup prior to every change, in personnel or conditions, sometimes in intervals of 30 minutes or less, and discuss the next steps, exactly who would do what, and what to do if things went sideways. There were no signoff sheets, no prewritten hazard analysis and no written instructions for managing risk. Just a group of men doing what they were trained to do while engineering new techniques to meet the unique challenges and risks each burning oil-well presented. The only rule I remember being told constantly was, "If anything happens while you are working at the wellhead

FIGURE 2.2
Rules-Based Safety Human Resource Comfort Zone

don't run. Slowly back out and we will cover you and get you out". Outside of that, every process was discussed at the moment on site prior to each task with one rule designed for recovery. I did not realise it then but this would be my first experience with "Post-Normal Science" being applied to safety management.

Prevention was built into the job much like pilots completing pre-flight checks not safety checks and spending countless hours in simulators practising various scenarios for recovery.

As a side note, Safety Boss went on to successfully cap more wells than any of the top four firefighting companies engaged in the project and did so with no serious injuries or lost time.

This experience is what would eventually lead me into the oil and gas industry overseas and my first offshore job with SECORP Industries as a medic/H_2S Tech on board a jack-up rig in the Persian Gulf. In turn, this led to joining Cliffs Drilling as a staff member in the newly created role of Safety Officer. I did not have much in the way of formal safety training or, for that matter, much of a job description, but what I did know is that offshore drilling crews knew what they were doing and did not need me to tell them how to do it safely.

As a young ambitious safety practitioner, working for Cliffs, one of my first assignments was to assemble safe work procedures, written by a handful of engineers in an office thousands of miles away, and from them create a binder of JSAs (Job Safety Analysis) or JHAs (Job Hazard Analysis). This binder would then be utilised by the driller on the rig floor to save time conducting pre-job safety meetings.

With the introduction of e-mail and the internet offshore, this job became even easier as it allowed us to share files between rigs and companies. After some time of being passed around and edited, via the copy–cut–paste technique, these files began to resemble procedures and the only portion that was ever seen by, or communicated to, the worker was the signoff sheet saying they had attended the pre-job meeting and were aware of the hazards and understood the procedures (rules) to manage the risk.

The book of rules became so big and cumbersome that the content became even more disconnected from the reality of the worksite.

In the early 2000s, following a wave of mergers in the offshore drilling sector starting in 1998, I found myself going from Cliff's Drilling, as a safety officer to R&B Falcon Drilling, as an RSTC (Rig Safety Training Coordinator). Different job name, same job description with a twist. Around this time, oil and gas companies were realising they could employ doctors from developing countries cheaper than advanced life support medics from the West while the drilling contractors discovered not all doctors are necessarily qualified to manage medical emergencies offshore. So aside from building rule books and hosting weekly flavour-of-the-day safety training, the expectation was that the RSTC, mostly US Nationally Registered Paramedics like myself, would be there to coach the new offshore doctors in a crisis.

Not only did my job description change but also my workplace transitioned from a jack-up platform to a drill ship, a much larger and more complex environment. The only thing that remained constant was the one-size-fits-all rules.

I am sharing this with you here as it explains a bit about my journey and paints the picture of how safety was being mismanaged in many industries as a function of rules with little thought of risk. My first day on a drill ship was much like my first day offshore with Cliffs whereby I, and others like me, represented a major risk, with nothing more than policy and procedure to manage us.

The Cliffs/R&B Falcon merger, and my subsequent job change, came a year after Reading & Bates merged with Falcon Drilling to become R&B Falcon. This merger was taking place the same year Transocean was merging with Sedco Forex to become Transocean Sedco Forex Inc. (TSF). Less than a year after this, in August of 2000, Transocean Sedco Forex acquired R&B Falcon to become the third largest oil field service company in the world and the largest offshore deep-water drilling contractor.

These events are relative because it demonstrates the vast number of policies and procedures workers were exposed to over a two-year period. Merging five-plus companies in under two years left workers confused, bitter and morally fatigued. To address this, the executives at Transocean promoted Lewis Senior, an experienced driller, tool-pusher and offshore installation manager (OIM), to the position of "Global Head of HSE".

His challenge was to merge multiple volumes of policy and procedure manuals, gathering dust on the shelves of four companies, to create a single safety management system (SMS) that crews could read, understand and implement.

In the International edition of the IOSH publication Principles of Health and Safety at Work[29] referenced earlier, Allan St John Holt states: *Workers need training to develop skills, to recognise the need to develop and comply with safe systems of work, and to report and correct unsafe conditions and practices.*

The new Transocean Sedco Forex certainly embraced the training and skills development of their rig crews. All rig personnel were made to attend initial training in the country in which they worked. This training was often attended by senior executives, including the CEO and his wife, creating an atmosphere of care and family.

These sessions were followed up in Houston, Texas, with "immersion training" for middle managers, supervisors and safety personnel. The immersion training emphasises the need to empower workers to be conscious of hazardous work environments and engage each other in better planning, appropriate protection and hazard identification. The system went so far as to identify each worker's personality type based on the platinum rule to treat others the way they need to be treated which was meant to form the basis for better communication and respectful work environments and, of course, the system included a pyramid.

This would be the first time that I was exposed to process-driven safety focused on risk versus rules and, while it was a welcomed departure from the volumes of rules-based manuals, it was not immune to the use of tools out of context and old-school methods.

In reference to Heinrich's three fundamentals, it embraced fundamental number one creating an active interest in health and safety but, due in large to comfort zones of the way we have always managed people, it fell prey to moral fatigue and moral disengagement among the crews, and as such, fundamentals two and three fell by the wayside while managers embraced quotas for collecting faulty information.

Risk

According to Angus Rhodes[30], the Global Project Manager at Ventiv Technology, the earliest concept of managing risk can be attributed to gaming. This is an appropriate place to start as the methods adopted by safety practitioners to assess risk in the field, borrowed from economic and financial risk management, amount to not much more than rolling bones or playing a chess match against luck.

There are many definitions of risk assessment. Nick W. Hurst[31] defines it as follows: *The study of decisions subject to uncertain consequences. It consists of Risk Estimation and Risk Evaluation.*

If you ever want to witness a feverish debate among safety practitioners, ask a group of them what "severity" and "likelihood" mean in terms of quantifying risk ($R = S \times L$), then sit back and enjoy a two-sided debate on a system that means absolutely nothing in terms of occupational health and safety.

The Health and Safety Executive (HSE)[32] in the UK defines them as the severity of harm and the likelihood of occurrence. By this, they mean how bad will it hurt and how likely is it to hurt that bad.

On the other hand, the Canadian Centre for Occupational Health and Safety (CCOHS)[33] advocates ($R = P \times S$), P being probability and S being severity. How probable, or likely, is an event to happen and if it happens how severe the outcome could be.

In other words, the first definition decides how bad you could get hurt falling off a ladder (paralyzed) against the chances of being paralyzed by falling. The second defines how likely you are to fall off the ladder and how bad falling off a ladder could be. One guesstimates the chances of getting hurt while the other guesstimates the chances of falling.

In the debate, there are two schools of thought. The first group, those who embrace the HSE definition, will say that you could die tripping on a rock but the chances of death by tripping on the same rock are very low and therefore not worth attention. This creates blind spots, and rather than removing the rock (the hazard), they calculate the likelihood of the consequence (death) as low. The second group, those who adhere to the CCOHS definition, will argue that you could die falling from a ladder, but if the likelihood of falling is very low (engineered controls), then the risk is low.

Both methods rely on the subjectivity of risk tolerance, and risk tolerance increases proportionately as competing values, money, time, etc., go up. Point in case, speeding through a school zone when late for an important meeting.

Then there is yet another equation for qualitative risk ranking.

$$\text{Risk} = \text{Probability} \times \text{Consequence} \times \text{Frequency}$$

This creates even more opportunities for misunderstanding. Is it the probability that the consequence will happen and out of how many times "one in a thousand", or is it the probability it will happen based on how often the task is performed?

In my experience, these methods are all flawed as they rely on a roll of the dice and a "what are the chances" or "it won't happen to me" gamble. Put into practice, it has been my experience, in both Front End Engineering Design (FEED) stages of major capital projects or nonroutine operations in the field, that probability and/or likelihood is the sacrificial lamb when faced with competing values like budget or work schedules. This is especially true in organisations that practice safety by incentive. As Terry McSween[34] has stated as follows: *The likelihood of getting a safety award is roughly the same as getting hurt.*

A great example of the 1/1,000 trap at the high end of the risk ranking process is the Deepwater Horizon disaster that killed 11 men and devastated the US southern gulf coast on the 20th of April 2010.

[Full Disclosure: As mentioned previously I was once a staff member at Transocean while it was still TSF. I have also consulted to BP as a health and safety advisor onboard a Transocean drill ship. Using this disaster as an example here, and in future chapters, is not a judgment or indictment of any kind on the men and women onboard the vessel or those ashore from any company or government organisation, nor is it meant to be disrespectful to the families of the men lost that day and those who continue to struggle with the aftermath. The event is used here as it is a great example of how the blind spots created by our seven bad habits can impact professionals and laymen alike. The opinions expressed here have the benefit of hindsight and I certainly do not have all the answers. My hope is simply to highlight improvement opportunities that may one day help to prevent the heartache and loss felt by the crew and their families.]

On April 21, 2010, I was working in West Texas facilitating a safety leadership programme for an oil and gas operator. My routine on these out-of-town jobs was to wake up early, press start on the hotel single-cup coffee maker, and turn on the TV to see what transpired while I slept. This was the first I would hear of the situation unfolding in the Gulf of Mexico onboard the Deepwater Horizon. For the remaining days, my co-facilitator and I took every opportunity to tune into updates on the situation. My motivation was driven by the fear that I might know some of the crew and their families from my time at TSF and perhaps a few of the names from my BP days.

One day, while driving to dinner, we were listening to a national radio programme where the then director of the US Department of Interior's Minerals Management Services (MMS), Elizabeth Birnbaum, was being interviewed regarding the drilling programme approval granted to BP for the Macondo well where the Deepwater Horizon had been drilling.

Federal regulations in the United States, and best practice, limit the placement of the seal which stops drilling fluid from contaminating casing cement (top plug) to no lower than 1,000 feet below the ocean floor. However, MMS had granted BP permission to install the top plug for this well at 3,300 feet below the mud line. The added advantage would be that any additional drilling mud, above the top plug, could be pumped to a workboat and reused elsewhere saving thousands upon thousands of dollars.

During the radio programme, the host asked Ms Birnbaum, something to the effect of "What process is used to assess the risks before approving these types of deviations?" To which she responded (paraphrased): *"We have a team of professional risk managers who conduct in-depth risk assessments prior to approving any drilling programs and, in the case of the Macondo Well, the chances of this happening were 1 in 1000".*

The problem with 1/1,000 being an acceptable level of risk is you cannot count 999 times and stop to prevent one out of a thousand outcomes. Yet we do this all the time both on major risk assessments like this and for daily risks with smaller potential outcomes in the field. An even bigger challenge is the bad habit of expecting untrained risk assessors to manage risk by utilising tools out of context to quantify their risk exposure assuming they all have a common risk tolerance.

We will return to this example at the bottom of the chapter, but for now, I want to explore the challenge hazard assessment plays in our out-of-context use of risk ranking.

Nick Hurst believes that risk, as a concept, is complex and that the confusion over quantifying it makes it more so while hazards he views as tangible. He writes as follows: *The concept of hazards, on the other hand, seem to cause less of a problem, and is variously described as 'objective' or 'real' as opposed to risk which is described as 'subjective' and 'constructed'*[35].

He defines hazards as: *The situation or object that in particular circumstances can lead to harm, i.e. has potential to cause harm.*

The first step in risk management is . . . you must see the hazard.

Everyone would agree that a poisonous snake is a hazard, but you cannot protect yourself from it if you do not see it. The risk involved in being around that snake on the other hand is subjective based on your knowledge (or perceived knowledge), skill (or perceived skill), experience with snakes and your behaviour around the snake.

The IOSH publication Principles of Health and Safety At Work[36] outlines three principles of accident prevention:

1. Use Techniques to identify and remove hazards.
2. Use techniques to assess and control remaining risks.
3. Use techniques to influence behaviours and attitudes.

The systemic failure many organisations face, when implementing hazard identification and risk control schemes, is the bad habit carried over from the rules-based safety management system where the policies, procedures and job safety analysis are often dictated by someone offsite and adapted to suit the task not the work environment. Checking boxes gives the false sense of security that all hazards have been identified and that the necessary controls (also decided from afar) are fit for purpose.

So, what happens when we take the three IOSH principles, listed above, and implement them via Heinrich's three fundamentals of creating and maintaining active interest in safety, fact finding, and corrective action based on the facts? While in theory this should serve to take the process out of the office and onto the job site, enabling workers to take control of their work environment, in practice it falls short.

The biggest reason for the systemic failure of hazard identification and risk management in the field is a combination of the tools we use out of context and a disconnect between the tools workers need versus the tool they are given.

The first disconnect is the continued use of the hierarchy of hazard controls, introduced in 1950 by the National Safety Council in the United States. The idea behind the model was to first consider eliminating the hazards; second to substitute hazardous processes or materials with non-hazardous ones; third to install engineered controls, such as guarding or fall restraint; fourth to develop administrative controls (policy/procedure/rules); and last to utilise personal protective equipment (PPE) such as safety glasses or fall arrest.

This concept works well when developing processes and planning projects at the management level; however, in the field the hierarchy is turned upside down as shown in Figure 2.3. Workers are traditionally policed on compliance with PPE requirements and administrative rules while most engineered controls and the ability to substitute materials or eliminate hazardous work environments remain in the hands of management.

So, if our techniques to identify and remove hazards, and assess and control remaining risks, as promoted by IOSH, are not working, we are left relying on IOSH principle number three – methods to influence behaviours and attitudes.

We will explore the challenges with behaviour-based safety in Bad Habit 5, but first, I would like to introduce the conceptual model of Zone Management, which aligns nicely with the IOSH principles and utilises Heinrich's three fundamentals for accident prevention.

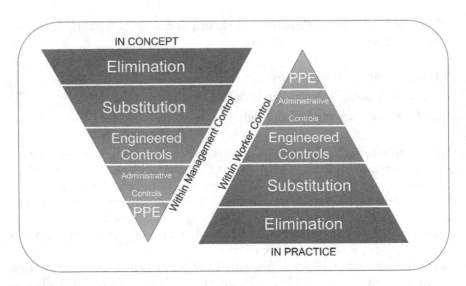

FIGURE 2.3
Hierarchy of Hazard Control as Conceived Versus as Practiced

Zone Management Theory

Zone Management Theory borrows elements from Rasmussen's risk management framework and Post-Normal Science and aligns them with Heinrich's three fundamentals of accident prevention.

Jens Rasmussen was a system safety, human factors and cognitive systems researcher at the Risø National Laboratory, Risø, Denmark. His work created important conceptual tools, such as the skills–rules–knowledge framework, and countless conceptual models, but it was concepts taken from his work on *Cognitive Systems Engineering* that we drew from when developing Zone Management Theory. *In a modern dynamic environment where discretionary decision making to a large degree is replacing routine tasks, definition of a correct or normal way of doing things is difficult.*

According to the paper *Working Near the Edge: A New Approach to Construction Safety*[37], traditional risk management programmes are built on the beliefs held over from the human era, specifically:

1. Rules and procedures can be developed which if followed will keep people safe.
2. Incidents happen because of worker error; that is, failure to follow the rules.
3. Reducing incidents will flow from improved motivation and training; that is, getting people to follow the rules.

The authors of the paper referenced above state:

> *We do not argue that worker motivation and training are unimportant, or doubt that people make mistakes and choices that lead to tragedy. But we do not believe that the worker centered cause and effect model, coupled with the violation of procedures, explains how incidents occur or provides the leverage required for further improvement. Additional reasons for moving beyond a worker centered model include:*
>
> * *Motivation and training have been a primary focus of efforts to improve productivity beginning at least since 1970. We now find that redesigning the production system using lean theory has greater impact.*
> * *Programs in general make us suspicious. More than a few corporate programs have been created to solve a problem without requiring deeper change. Productivity programs were tried extensively beginning in the 1970's but made only modest gains and rarely caused a fundamental shift in the way work was done. It has been said that all programmatic fixes to organizational problems eventually pass away; they either cause a change in the fundamental practice they were chartered to affect or they become obviously impotent and are cancelled. We expect safety programs will continue because they are almost required by regulation, and no better approach is yet apparent.*
> * *There appears to be a more powerful theory and approach.*

They believe that the framework proposed by Jens Rasmussen in Cognitive Systems Engineering offers a broader and more powerful view of the relationship between the individual and the work environment. Without going into detail on Rasmussen's theories, the basic belief is that work migrates away from both the organisational boundary, fear of economic consequence, and the individual boundary, a distaste for excessive control. He believed that accident causation is about a loss of control which occurs when the work migrates to a boundary of functionally acceptable behaviour. In short, a combination of job-scope-creep and normalisation of deviance.

As shown in Figure 2.4 when we adapt this to our conceptual model of Zone Management Theory, we create three work zones:

* Safe Zone
* Hazard Control Zone and
* Loss Control Zone

Traditional safety regulations and management practices are directed at keeping the workers in a safe zone. As mentioned above, no amount of regulatory or supervisory efforts will contain work to the safe zone, workers will migrate to the hazard zone for many reasons. Rasmussen suggests that enlarging the safe zone through proper planning of operations is required to ensure safe work both in and out of the safe zone.

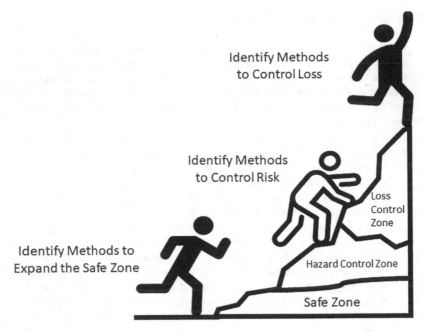

Identify Methods
to Control Loss

Identify Methods
to Control Risk

Loss
Control
Zone

Identify Methods to
Expand the Safe Zone

Hazard Control Zone

Safe Zone

© M. Ritchie 2023

FIGURE 2.4
Work Zone Theory

The white paper "Working on the Edge" by Howell, Ballard, Abdelhamid, and Mitropoulos (referenced above) when speaking of the original Rasmussen model wrote:

> *This model challenges current safety program practice on a number of fronts, including the concept of error based on standard procedure. . . . But the fundamental difference flows from the recognition that both individual tendencies and organizational factors push people to work in risky circumstance. Recognizing the inexorability of the forces at play, it appears necessary to develop coping behavior at the edge of control. This challenges the notion that workers can be kept inside the safe zone and should never enter the danger zone where loss of control is possible.*

They go on to note that the approach recognises that people will adapt to zone changes and therefore supporting them to develop and utilise judgement is far more effective than holding to the age-old notion that rules and procedures can be developed which if followed will keep people safe.

Rasmussen believed that the worker is the best person to judge the boundaries of safe work. The first step in Zone Management Theory is taken from the Howell et al. white paper, which suggests training workers so they can answer these four questions.

- Where are you – in what zone?
- What is the risk or hazard you now face?
- What can be done to prevent releasing the hazard?
- What can be done to reduce harm should the hazard be released?

The implementation of zone recognition versus risk assessment removes the rolling of the dice and subjective risk tolerance and replaces it with identifiable boundaries, which can be managed in real time. Compliance does not equal safe operations access to relevant choices does.

We must also rethink incident investigation and analysis to consider the design of the work in relation to these concepts, thereby achieving Heinrich's third fundamental of accident prevention, corrective actions based on facts.

Rasmussen believed that the worker is the best person to judge the boundaries of safe work. Rather than forcing compliance with rules intended to keep them in the safe zone, he suggested training workers to identify the boundaries and give them access to choices based on operational realities. Having the tools to make decisions in real time based on facts, far-out ways calculating what if probabilities on a sliding scale of tolerance impacted by outside pressures such as money and time. This is where we apply the concept of post-normal science.

Post-normal science (PNS) was developed in the 1990s by Silvio Funtowicz and Jerome R. Ravetz. According to Funtowicz[38], it is intended for use as a method for problem-solving when facts are uncertain, values are disputed, stakes are high and/or the need for urgent decisions.

When defining examples of its intended use he writes:

> *A good example of a problem requiring post-normal science is the actions that need to be taken to mitigate the effects of sea level rise consequent on global climate change. . . . The COVID-19 pandemic is another instance of a post-normal science problem. The behaviour of the current and emerging variants of the virus is uncertain, the values of socially intrusive remedies are in dispute, and obviously stakes are high and decisions urgent.*

While safety management may seem a far distance from climate change or a global pandemic, the reality is that nothing better describes the current state of the safety profession than uncertain facts, disputed values, high stakes and the need (or perceived need) for urgent decisions. With respect to conservation, Falko Buschke, Emily Botts and Samuel Sinclair[39] promote post-normal science as filling the space between research, policy and implementation. Much like conservation efforts, safety management suffers from a "research-implementation gap" making the concept of post-normal safety worth exploring.

Silvio Funtowicz explains Figure 2.5 as:

> *When systems uncertainties or decision stakes are small, we are in the realm of 'normal' science, where expertise is fully effective. When either systems uncertainties or decision stakes rise then skill, judgement and sometimes even courage are required. This is the realm of professional consultancy. And when either or both systems uncertainties or decision stakes are high this is the realm of post-normal science.*

Applying this explanation to zone management theory in safety when decision stakes are small and system uncertainty is low, we are in the safe zone. The safety management systems in place are known to be effective and are founded on our professional practices.

When decision stakes increase and/or the system uncertainty increases, we enter into the professional consultation zone. Professional consultation includes everything from involving supervisors or managers in the decision-making to outside consultancy in the form of subject matter experts. Regardless of the level of consultancy, the migration from the safe zone to the hazard control zone signals the need to step back and identify the underlying elements pushing the work uphill.

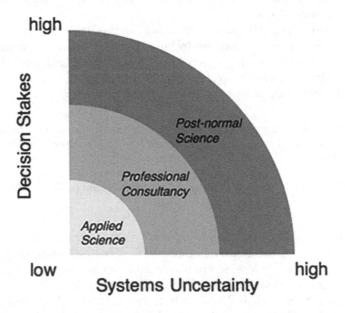

FIGURE 2.5
Post-Normal Science

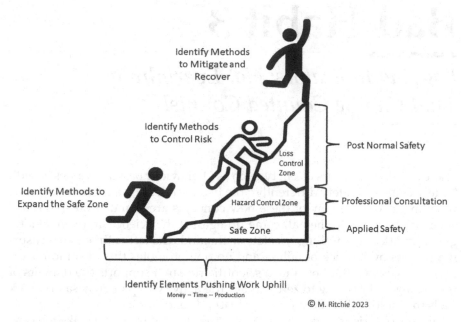

Identify Methods
to Mitigate and
Recover

Identify Methods
to Control Risk

Post Normal Safety

Loss
Control
Zone

Identify Methods to
Expand the Safe Zone

Hazard Control Zone

Professional Consultation

Safe Zone

Applied Safety

Identify Elements Pushing Work Uphill
Money – Time –- Production

© M. Ritchie 2023

FIGURE 2.6
Zone Management Theory

At the point where either the decision stakes are critical and/or system uncertainty cannot be resolved by consultation, we have drifted into the loss control zone signalling the need for a full stop to explore both choices available to either expand the safe zone or implement controls putting the work back in the hazard control zone. It is also crucial at this stage to identify options to mitigate and recover from the possible loss.

Consideration for mitigation and recovery should focus on the financial impacts as much as they do on physical loss. Simply focusing on physical loss drags us back to the bad habit of severity versus probability. Financial impacts must be focused on realising the cost of recovering from the unplanned release of energy and not on the cost to mitigate it or the money saved by accepting it.

Post-normal safety in the loss control zone requires identifying and managing complicated uncertainties in knowledge and cognitive bias and recognising the legitimacy of opposing perspectives and ways of knowing.

In the remaining chapters, we will explore bad habits that must be addressed to combat the research–implementation gap that continues to plague the profession as we enter the holistic era in safety.

Bad Habit 3

Lagging Indicators: Not Everything That Can Be Counted Counts!

There are many flavours of safety, such as "behaviour based", "values based", "safety culture", "safety differently"; the list is endless. You name it; someone is selling it. While many of these new concepts are founded on sound science and have merit, they all come up against the science–implementation gap. Why? Because nothing in safety withstands the test of time like measuring success by the lack of failure and no one promotes this idea more than regulators. No matter how much scientific research supports the dangers of rewarding and reacting to zero targets, the activity trap of busy safety work fuels the multimillion-dollar zero industry. Money for nothing.

Like my old friend and mentor, Alan Quilley, used to say, "It's like measuring fitness by the absence of sickness".

Judy Agnew and Aubrey Daniels[4], referenced in Chapter 1, said it best: *Sophisticated companies that use only the latest scientific information and technologies from chemistry, physics, engineering, and biology, use so-called common sense, myth, and downright faulty information to manage the behaviour of their employees.*

And no such faulty information tops the list of bad habits in safety management like the total recordable incident rate (TRIR) and the total recordable incident frequency (TRIF).

While they certainly serve a purpose for regulators, as alarm bells for non-compliance, and insurers, for setting experience ratings for industry premiums, they are terribly misused and misunderstood at the ground level of safety management.

Many years ago, I was working offshore in Africa on a jack-up drilling rig. After establishing a good relationship with the Offshore Installation Manager (OIM), I was able to convince him that he needed to lead the weekly safety meetings. I told him that no one was really interested in what I had to say, but as our leader, his words would be heard and motivate crews to participate.

Reluctant to take on this role, he asked what was entailed and I went over the various must-do and a few good-to-do agenda items and told him that I would help with the preparation. After a combination of coaching and begging, I finally convinced him it was an opportunity for him to demonstrate his commitment to safety onboard the rig.

One of the "must do's" for our weekly safety meeting was to review the year-to-date safety statistics which included the TRIR. This was an item the

DOI: 10.1201/9781003404958-3

rig crews, including the OIM, had heard referenced weekly for as long as they had been with the company.

On the OIM's first day leading a meeting, I had him all set up with a PowerPoint presentation and the crew eagerly awaited his debut at the front of the room. After a bit of a shaky start, he began to show some passion for the messages he was conveying. Then, he got to the slide with the year-to-date TRIR, and he said: *Guys the rig's TRIR to date is one point five but I know if we all pull together, we can get that up to two.*

That is when I realised that he, and probably most of the crew, had no concept of what these weekly stats were and how they were used. I can also say with a high level of certainty that they did not care. What they cared about was getting that quarterly safety bonus and the accompanying hats, jackets and keychains. And all they had to do to get that was keep quiet when there was an incident.

We will talk more about the dangers of "zero" incidents, but first, I want to explore the TRIR and TRIF and examine their benefits while unearthing some misconceptions.

Perhaps the crew of the TODCO 185, off the coast of West Africa back in 2005, can be forgiven for not understanding the meaning or relevance of the TRIR. It seems that even the regulators cannot agree on how to calculate incident rates, what they represent or how they should be used.

In the UK, the Health and Safety Executive (HSE)[40] defines the TRIR as representing a rate per 100,000 employees and calculate this rate by dividing the number of injuries in a fiscal year by the average number of employees then multiplying this by 100,000

$$\frac{\text{Number of reportable injuries in the fiscal year}}{\text{Average number of employees}} \times 100,000$$

In the United States, the Occupational Safety and Health Administration (OSHA) defines the TRIR as representing the rate per 100 workers. They calculate it as the number of incidents in a year multiplied by 200,000 divided by the total number of people hours worked in that year.

$$\frac{\text{Number of reportable incidents in the year}}{\text{People hours Worked}} \times 200,000$$

200,000 is used as the multiplier in this case as it represents the average hours worked by 100 people in a year. (100 workers × 40 hours × 50 weeks)

To add to the confusion, many organisations in Europe and other jurisdictions around the world like to calculate frequency rather than rate. They do this by counting hours worked versus people working. The logic is that counting hours rather than people is more accurate than counting employees who may or may not be full time.

$$\frac{\text{Number of reportable injuries in a specific period}}{\text{Total hours worked in that period}} \times 1,000,000$$

The frequency in this case is based on the number of incidents per million hours worked. Many organisations and regulators will adjust the multiplier down from one million depending on their situation or preference. The challenge this presents is a distorted picture when comparing frequencies across organisations or industries. One may be a rate of 5 per million hours worked while the other is 2 per 100,000 hours worked, and so on.

The HSE explains the use of their TRIR (per 100,000 people) and frequency (per million hours worked) as: *Accident incidence and frequency rates provide a means of measuring safety performance over time and comparing it with accident statistics published by external sources, such as HSE.*

The same "safety performance" we celebrated with crews offshore while they compared their weekly battle scars and solicit tensor bandages, anti-inflammatories, and Band-Aids from the medic.

Again, quoting Alan Quilley[41]: *We're bombarded by information some of it statistically sound . . . some not so much. Some of it presented in very misleading ways to try and convince you of a particular bias.*

In this instance, Al was referring to the bombardment of statistics in relation to the COVID pandemic, but in the world of occupational safety, nothing demonstrates the power of statistic bias like the statement made by Transocean to shareholders in 2010, following the Deepwater Horizon disaster, boasting their best safety record to date. Company executives were quoted as saying:

> *Notwithstanding the tragic loss of life in the Gulf of Mexico, we achieved an exemplary statistical safety record as measured by our total recordable incident rate ("TRIR"). . . . As measured by these standards, we recorded the best year in safety performance in our Company's history, which is a reflection on our commitment to achieving an incident free environment, all the time, everywhere.[42]*

It would appear that when you multiply 11 deaths by 200,000 and divide that by millions of people hours, the number is insignificant to shareholders and proves your "best year in safety".

Regardless of the method used, people, hours, etc., the reality is that the collection and comparison of rates or frequencies are subject to the accuracy of reporting, what constitutes a "reportable" or "recordable" incident and the purpose for collecting them. Yes, there is value for an insurance company or worker's compensation board to track incident rates by industry to calculate the experience rating of the industry and thereby calculate insurance premiums. But aside from providing a spitball figure for the actuaries, tracking rates and frequencies based on faulty information does nothing for the worker in the field but create moral fatigue fuelled by the organisation's focus on last year's news while real-time hazards and risks go unmentioned or unattended. While the numbers provide a very subjective "how many" they fail to identify the "why's, how's, where's and what's".

The dangers of chasing zero have been demonstrated time and time again by many researchers and writers, yet we somehow lack the political will to change. This could be the best example of the research–implementation gap mentioned in Chapter 2. I am not suggesting insurers and regulator abandon their calculations to gauge potential financial impacts, but this activity would be best left between the safety and human resource managers and regulators. The only time the workforce should be involved is when the organisation or regulator celebrate and reward reporting.

What we have learned time and time again from major incidents around the world is that the consequences of focusing on personal safety metrics, like lost time and TRIR, while neglecting the measurement of real-time challenges and competing values, create blind spots focusing everyone's attention on high-frequency low-consequence events. Both BP and Transocean had high-level management onboard the Deepwater Horizon the day of the explosion and they were there to celebrate 7 years of no lost time, which everyone onboard knew meant 7 years of nonreporting. This was during a time when the rig crew was drifting from the hazard control zone to the loss control zone and the professional consultation was being dragged back to the applied safety comfort zone when it should have been shifting to a postnormal safety approach.

It is this target of an "incident free environment" that creates operational secrets and the "catch us if you can" mindset that middle managers and crews gravitate to in the quest for bonuses and BBQs.

So where did this concept of zero come from? Why is it that regulators and senior executives and managers fear hearing the truth and punish those who speak it? And whatever made safety practitioners think that a lack of reports equalled a safe work environment?

It has been said that Pyramids are where obsolete cultures bury their dead leaders. Yet no one enjoys a good pyramid like the safety practitioner. Much like the cycle of Plan–Do–Check–Act and calculating risk based on playing the odds, pyramids and triangles have been reworked, repackaged, resold and taught out of context for nearly 100 years.

When exploring lagging indicators, it is important to look at the complete picture to see what, if anything, is worth keeping from the bad habit of looking backwards to see what is in front. Ironically, the best place to start this journey is to look back.

Statistical accident research, focusing on outcomes, did not start with Heinrich but first appeared in the early 1900s around the start of the human era in safety management, long after the industrial revolution. The focus on outcomes rather than prevention was due to commonly held beliefs that injured workers were responsible for their injuries, so cost, not causation, became the driving force behind the early studies.

In 1907, The Russell Sage Foundation, in the United States, sponsored a sociological examination of urban conditions in Pittsburgh, Pennsylvania. Volume II of the six-volume "Pittsburgh Survey"[43] was written by Crystal Eastman, a labour lawyer, suffragist and journalist. Her volume examined work injuries and the legal compensation that workers were offered.

This is considered by many to be the first sociological study of a particular demographic in relation to occupational safety and one of the factors that would lead to the various states in the US adopting worker's compensation schemes.

Setting the stage for studies to come the focus was on accident outcomes rather than causation as incident investigations were limited to coroners' inquests following a fatality and human error or act of god was the standard predetermined cause.

Likewise, her study of workplace injury was limited to injuries that required hospital treatment and therefore had traceable medical records. The sad fact of early 1900's America was that most injured workers could not afford either the medical treatment or lost days wages, and the lack of a formal workers' compensation scheme in Pennsylvania at the time meant most injuries were not recorded.

It is worth pausing here for a moment and taking a deeper dive into her study as it is not so much the numbers she presented, but her interpretation and personal conviction which impacted safety management practices for the next 100 years.

Describing her study demographic, she wrote:

> *Allegheny County, which roughly corresponds with the famous Pittsburgh "Steel District," has a population of 1,000,000, of whom 250,000 are wage-earners. Seventy thousand in the steel mills, 20,000 in the mines, 50,000 on the railroads. These are the great employment groups in Allegheny County; they are also the great accident groups.*

Among Eastman's findings, she reported that, from July 1, 1906, to June 30, 1907, 526 men were killed by work accidents in Allegheny County, Pennsylvania. She also reported that during a three-month period, April, May and June, of that same year, the hospitals of the county received over 509 workplace injuries requiring medical attention.

Her injury rates focused exclusively on life-altering injuries and the debilitating conditions experienced by the injured, their families, society and the economy. In today's terms, the focus was how incidents at the top of Heinrich's triangle (Figure 3.1) impacted society.

These injuries were divided into two groups, slight permanent injury and serious permanent injury. In describing what constituted serious injury, she wrote:

> *Ninety-one sustained what is here called a slight permanent injury; for instance, a lame leg, arm, foot, hand, or back, not serious enough to disable a man, the loss of a finger, slight impairment of sight or hearing, and the like. Seventy-six men (25.5 per cent) suffered a serious permanent injury. Lest there should be doubt as to what is meant here by "serious," it will be better to state exactly what these injuries were. Seven men lost a leg, sixteen men were hopelessly crippled in one or both legs, one lost a foot, two lost half a foot, five lost an arm, three lost a hand, ten lost two or more fingers, two were left with crippled left arms, three with crippled right arms, and two with two useless arms. Eleven lost an eye, and three others had the sight of both eyes damaged. Two men have crippled backs, two received internal injuries, one is partially paralyzed, one feebleminded, and two are stricken with the weakness of old age while still in their prime. Finally, three men suffer from a combination of permanent injuries. One of these has a rupture and a crippled foot; another a crippled left leg, and the right foot gone; the third has lost an arm and leg. These 76 are the wrecks of 294 hospital cases.*

Explaining her priority to focus on serious injury over fatality she wrote as follows:

> *Five hundred and twenty-six men dead does not necessarily mean 526 human tragedies. We all know men who would give more happiness by dying than they gave by living. But 500 men mutilated-here there can be no doubt.*

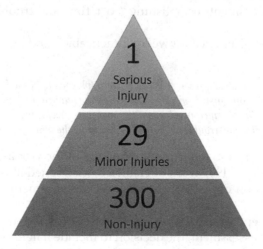

FIGURE 3.1
The Heinrich Triangle

Her take on the impact of workplace death versus long-term injury may have been influenced by the compensation aspect of her study. In financial terms, a workplace fatality was soon forgotten, and families moved on, mostly by sending another son into the plant to make up lost family income. On the other hand, the sick and lame remained in the public eye and continued to stress society and the economy.

Pennsylvania would not introduce mandatory workers' compensation until 1915, 5 years after the publication of Eastman's study, and this drew the actuaries and insurance industries' attention to calculating the costs of underwriting workplace accidents. The entry of financial risk management to the safety profession added to the engineering initiatives and the frequency of incident outcomes became the formula to assess cost and set insurance premiums.

In Chapter 2, we talked about the calculations used to rank risk. Likelihood × Severity, Probability × Consequence, etc. No matter how you calculate it, as mentioned in Chapter 2, when up against production quotas and cost overruns, likelihood and probability always win. "It will never happen to me . . . I'm faster, stronger, smarter . . . Chances are one in a thousand . . . etc.".

From a cost perspective, probability plays a much larger role which is where we get terms like ALARP (As Low As Reasonably Practicable). Many organisations often confuse "practicable" with "possible", but the two terms are very different.

Arising out of a legal case in the UK in 1949, Edwards v. National Coal Board, the case of a worker killed by a road collapse, the National Coal Board argued that it would be too expensive to shore up every roadway in every mine. This led to the ruling by Court of Appeals Judge Cyril Asquith that not all roadways needed shoring up only those determined to be "practicable" based on risk assessments establishing "cost, time and trouble against likely outcomes"[44].

Describing the difference between "practicable" and "possible", Asquith wrote as follows:

> *Reasonably practicable is a narrower term than 'physically possible' and implies that a computation must be made . . . in which the quantum of risk is placed in one scale and the sacrifice involved in the measures necessary for averting the risk (whether in time, trouble or money) is placed in the other.*

This created the need to determine frequency in the risk assessment which drew attention to the 1931 work of Heinrich, mentioned above, and justified further examination of his calculations and measurements which continue to this day.

While the concept of ALARP may seem better explored in Chapter 2, where we explored risk assessment, the decision to include it here is twofold. First, it highlights the original use of the research, which was to determine cost not workplace hazards and risk, and second, the term ALARP means absolutely nothing past a flimsy due diligence defence following a major incident and does nothing to promote worker safety. Most often, it helps create a justification for taking the low road.

However, when exploring the measurement of both lagging and leading indicators the triangles are worth examining before they are dismissed entirely.

Herbert William Heinrich was Assistant Superintendent of the Engineering and Inspection Division at Travelers Insurance Company. His work studying incident frequency against severity was first introduced in 1931 as part of his book *Industrial Accident Prevention: A Scientific Approach*[45], which has been referenced in Chapter 2. The intent of this portion of his study was to deliver quantifiable numbers to assist insurers in calculating financial risk based on the frequency of severity to perhaps help calculate probability.

Among his findings, Heinrich reported that for every 330 "similar events" 1 will result in major injury, 29 will result in minor injury and 300 will result in no injury. (Figure 3.1)

The original concept was very straightforward and useful when applied to insurance costs. If the cost of insuring a fleet of 330 vehicles is greater than the cost of minor repairs on 29 and writing off one, then it is good business to self-insure by budgeting for the cost. This assumes all 330 vehicles drive the same routes under the same conditions.

In his writings regarding the 1–29–300, Heinrich stated that the ratio related to 330 occurrences of the same kind. In short, because the events are similar, they all had the potential to cause a major injury, but in 329 instances, they did not because of luck or some other circumstance. As an example, of 329 cases of a roofer falling off a roof, 300 will walk away, 29 will limp away and one will be carried away.

The first mistake when applying this concept to incident prevention is the assumption that all incidents have a common cause. The notion that if a task is performed 330 times, one of those times will result in a serious incident is not only ridiculous but not at all what Heinrich was proposing. In his writings, he clearly stated that similar events may or may not have similar outcomes.

So how then did the safety profession get to a place where frequency equals severity and serious incidents have a time stamp based on "it's only a matter of time"?

Perhaps some of the confusion can be attributed to Frank E. Bird, another insurance researcher who worked for the Insurance Company of North America.

In the late 1960s and early 1970s, Bird headed a study which analysed more than 1.7 million accidents reported by 297 cooperating companies. These companies represented 21 different industrial groups, employing 1.7 million employees who worked over three billion hours during the period analysed[46].

Bird's original findings suggested that the ratio between fatal accidents, serious injuries, minor injuries and no injury incidents was 1-10-30-600. Keep in mind that Bird analysed ratios based on incidents reported to the insurance company with no relationship to chronological sequence or frequency of tasks.

The key phrase here is "reported incidents" as his data were based solely on accidents that were reported to the insurance company, which, no doubt, were very different from the real number of accidents and incidents taking place at the worksites.

In the book Loss Control Management: Practical Loss Control Leadership[47], Bird et al. wrote:

> *More important than ratios that usually vary with every organization is the fact that there are many opportunities afforded by the more frequent, less serious uncontrolled events to take preventative action before the major loss ever occurs.*

This is part of the reason we focus on high-frequency low-consequence events as was the case with the Deepwater Horizon. This also assumes that the less serious events share the same causation as those with more serious outcomes.

Responding to mounting industry criticism and challenges aimed at the triangle and its ever-changing outcomes, the Health and Safety Executives Accident Prevention Advisory Unit (APAU) in the UK published yet another study[48].

This study listed the top of the pyramid as "over three days lost time", which is a far cry from Crystal Eastman's definition of "serious injury" or as counted by Heinrich in 1931. The HSE triangle reported a ratio of 1 major or over three-day lost-time injury to every seven minor injuries to every 189 noninjury accidents.

What constituted a "minor injury" is not clear and, realistically, a sprained ankle could lead to more than three-day lost time depending on the physician.

This certainly is not a criticism of the HSE, Bird et al and does not imply that good things did not come of their work. In the next chapter, we will look at leading indicators and come full circle back to Heinrich and Bird to find those bits which can help take our eyes off the rear-view mirror. However, Herbert William Heinrich's original findings on accident outcome frequency have been challenged, changed, updated and turned into icebergs, and not unlike Deming before him, the newer versions, while not his, are almost always referred to as Heinrich's Triangle.

It is important here to pause and restate the simple fact that Heinrich's work was intended for use by insurance actuaries and underwriters. Also worth noting is his findings represented a ratio between outcomes from similar events, unlike many of the newer results credited to his name. His original work was simply meant to help calculate the potential cost of injury or property loss based on the frequency of incidents of the same type, the key focus being on cost not on prevention.

Perhaps this misguided love affair with triangles is founded in the reality that, like the PDCA model and the practice of managing risk with rules, the triangle is misunderstood and applied out of context.

Counting Crows

One for sorrow, Two for mirth, Three for a funeral, and Four for birth.

The tradition of foretelling the future by the flight of birds is known as *Augury* and is about as accurate as controlling workplace hazards by celebrating events that did not happen.

Several years ago, I was presenting at a professional development conference where a well-seasoned colleague was also presenting. Their presentation was a case study on a retail client and their journey to "zero".

As part of the presentation, the presenter said something to the effect of:

> *then they had a fatality on site. It was my first time dealing with a fatality but given Heinrich's triangle I shouldn't have been surprised. When you manage safety long enough it's just a matter of time in any safety professional's career before they will have to deal with a fatality.*

This statement floored me. How could this presenter, a well-qualified and experienced safety consultant who has held high-level positions in two national associations, interpret Heinrich or Bird's triangles as *"just a matter of time"*?

While I pondered what could possibly lead them to misunderstand Heinrich's original work, the answer came to me in the form of the presenter's answer to a rhetorical question . . . *"so what did I do? . . . I went back to the client and started again because Plan-Do-Check-Act is what we do"*.

The combination of the presenter's lack of understanding of an age-old insurance tool coupled with the comfort zone of doing the same thing over and hoping for a different result (PDCA) made me realise that perhaps the recent focus on debunking Heinrich and Bird is not because they do not have value but more so because of the way they are taught and used out of context.

There seem to be two concepts at play in terms of using safety triangles, pyramids or icebergs.

One is the notion shared by my colleague referenced above. "It is only a matter of time". In this mindset, the focus becomes categorising incidents, so they can be counted and quantified to measure success or failure. This is the lagging indicator approach regulators require and one that is utilised by large organisations when qualifying contractors or seeking certification in various safety schemes. It is what drives our attention to high-frequency low-impact events.

The other is the notion that if we focus on preventing incidents at the bottom of the pyramid, we will stop more serious incidents at the top. This is the leading indicator approach we will explore in Chapter 4.

In 1910, Crystal Eastman, addressing why the injury portion of her study for the Pittsburgh Survey, was conducted over a three-month period as opposed to the one-year period allocated to counting fatalities, wrote as follows:

> *It is impossible to state the total number of injuries during that quarter, because there is no available record except of cases received at the hospitals. But even were an accurate estimate of the number of injuries in a year possible, it would be of little value. A scratched finger and a lost leg can not be added together if you look for a useful truth in the sum.*

As far back as 113 years ago, Eastman, considered by many to be an early safety research pioneer recognised and documented the flaws in trying to extrapolate truthful data comparing outcomes to causation such as "how many scratched fingers does it take to lose a leg?" Yet in 2023, despite overwhelming research and data, regulators, such as the HSE and others, continue to invest in studies designed to support their low-hanging fruit of zero.

As mentioned previously, the first step is to leave the lagging indicators to the actuaries and start celebrating lessons learned from open and honest reporting.

So far in the book we have discussed the activity trap of PDCA and the need to stop running in circles and the dangers of managing safety and risk with rules, and here, we added the bad habit of collecting faulty information to decide on the Act in the PDCA cycle. Next, we will discuss the bad habit of developing plans derived from implementing leading indicators out of context and the comfort zone of fortune telling.

Bad Habit 4

Leading Indicators: Not Everything That Counts Can Be Counted!

In Chapter 2, we looked at safety management systems and measuring risk, and in Chapter 3, we discussed lagging indicators. The relationship between the two is what often fuels the activity trap described in Chapter 1.

To summarise the connections between Bad Habit 1, Bad Habit 2 and Bad Habit 3, we see the comfort zones play out as:

1. We plan for zero incidents focused on the past.
2. We dictate policy and procedure in the form of cardinal, life-saving rules, or some other cleverly named zero tolerance "don't get caught" scheme.
3. We check for compliance and count incident reports or the lack thereof.
4. We react (punish or restate zero policy) or we reward (safety bonus, BBQ and swag).
5. Lastly, we jump right past the plan to do more compliance and bonus-driven activities in search of a magic bullet that will support the original plan.

But what if that plan was flawed and unattainable? This is where leading indicators should come into play; however, because they are so often applied out of context and misused or misunderstood, they fail to provide a factual assessment of the operational realities and the current work environments.

In the remaining chapters, we will highlight opportunities to break the seven bad habits by simply adjusting how we utilise old and new concepts and tools proactively. Chapter by chapter, we will add another link that, when combined with a collaborative approach, supports Heinrich's first principle to "create and maintain an active interest in safety" (e.g., CARE), and provides a systemic method to support fundamental number two, "fact finding" (Zone Management Theory), which will allow us to develop corrective action based on facts (Heinrich's third fundamental of accident prevention). Sounds easy, unless our focus is firmly planted on desired outcomes as the regulators and compliance systems sales team tell us it has to be.

DOI: 10.1201/9781003404958-4

Many students studying OHS today are being taught that measurement is essential to maintaining and improving health and safety outcomes. In the publication "Successful Health and Safety Management"[49], the Health and Safety Executive (HSE) in the UK outlines two ways to generate information and measure performance. The first is via active systems and the second reactive systems.

They describe these as:

- Active systems which monitor the achievement of plans and the extent of compliance with standards and
- Reactive systems which monitor accidents, ill health, and incidents.

In both their examples, we are looking at lagging indicators. The publication goes on the say that: The key to effective monitoring is the quality of the plans, performance standards and specifications which have been established.

It is directives such as these that keep us trapped in Bad Habit 2, rules-based safety management. The key to effective monitoring is NOT the quality of the plans, standards or specifications. The key to effective monitoring is fundamental number one. The creation and maintenance of active interest in safety, the focus being on the reality or effectiveness of the plans, standards, and specifications.

If the plan is simply to report zero incidents, with the intent being to achieve a TRIR of zero so that the company is compliant with a pre-bid specification for a misguided contractor management system or report card you not only set yourself up to fail but you create the habit of not reporting and squash any chance of identifying corrective actions based on fact.

The directive given by the HSE is that organisations need to have procedures to allow them to collect the information to adequately investigate the cause of substandard performance. The irony with this statement is that, while the publication attempts to paint the active systems as leading indicators, the examples used to describe various forms of active monitoring include the following:

- Routine procedures used to monitor objectives in the form of quarterly or monthly reports (yesterday's news).
- Periodic examination of documents to check that systems related to the promotion of the health and safety culture are complied with (check for compliance).
- Systematic inspections of premises, plant and equipment by supervisors, maintenance staff, management, safety representatives, and staff to ensure continued use of workplace precautions (focused again on compliance rather than fit-for-purpose).
- Systematic direct observation of work and behaviour by first-line supervisors to assess compliance (rules-based blame the worker compliance).

- The operation of audit systems (time spent preparing to look good on audit day while the workers struggle with worksite realities).
- Consideration of regular reports on safety and health performance by the board of directors (your concerns will be addressed at the next board meeting).

The problem is not the specific activities but the way in which we manage and measure these initiatives combined with the starting point that defines them.

Programmes promoted to be leading indicators, such as observation cards, get processed as lagging indicators and represent an example of the "faulty information" being collected driven by the measurement of compliance in the PDCA activity trap. When we couple this with things like contractor prequalification tools or schemes intended to reward zero, at both the worker and organisational levels, we have the perfect storm to create fiction over fact. Regardless of the statistics we develop or how we spin them in our favour, we remain driven by faulty information.

To quote Josiah Stamp[50], a British civil servant, industrialist, economist, statistician and banker in the early 1990s:

> *The government are very keen on amassing statistics. They collect them, add them, raise them to the nth power, take the cube root and prepare wonderful diagrams. But you must never forget that every one of these figures comes in the first instance from the village watchman, who just puts down what he damn well pleases.*

At his point, it might be helpful to establish what, in my experience, a leading indicator should be and how it differs from historic information.

As noted in Chapter 3, a lagging indicator is reactive and reports on loss or noncompliance. In short, they report yesterday's news.

In contrast, a leading indicator is preventative and reports current news that could impact tomorrow, not tell the future.

The difference in application is demonstrated in Figure 4.1 below. Lagging indicators are based on desired outcomes, like TRIR, that are a measurement of failures. Leading indicators, when utilised proactively, are focused on real-time operations, which allow us to implement corrective actions based on facts.

As Figure 4.1 illustrates when we drive our efforts by desired outcomes, usually zero TRIR etc., our focus gets consumed by the numbers and this creates blind spots resulting in a reality of failure. However, when we look forward and focus on operational realities we are able to assess what lies ahead in terms of workflow. Our desired outcomes become task-oriented and we succeed in extending the safe zone, knowing when we have entered a hazard zone and the required controls in place and, most important, identifying when and why we have drifted into the loss-control zone signalling us to step back and engage the team.

It could be argued that developing "systematic inspections of premises, plant and equipment by supervisors, maintenance staff, management, safety

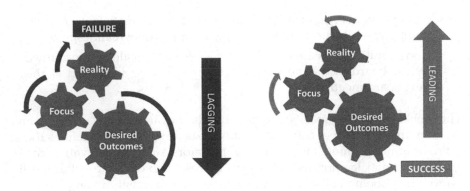

FIGURE 4.1
Lagging Versus Leading Indicator

representatives and staff to ensure continued use of workplace precautions" as outlined by the HSE is proactive. There is certainly merit in this so long as the process supports the creation of an active interest on the part of the work teams. Far too often these come in the form of the once-a-month committee meetings, executive site visits and VIP tours or audits with little or no input or output from or for the workers.

Implementing preventative maintenance programmes as routine job tasks (oil changed after xyz hours of operation) or verification (the space survey revealed x amount of tools recovered and returned to the tool crib following routine maintenance) is essential yes but they are not and should not be a function of a safety manager or, worse, safety officer. These are things we must build into trades and professional training.

Many of the directives suggested by the HSE, IOSH, et al., come from a place of good work practices but somehow found their way into policing safety compliance.

I have a friend who is a retired airline pilot, and he once told me that in his entire career, flying transatlantic flights, he never once did a safety check. What he did prior to every take-off and following every landing was a pre-flight and post-flight check of the aircraft. There were no safety policies attached; it was just the way he and all pilots are taught to fly. Can you imagine sitting on an airliner and hearing the pilot say "ladies and gentleman we are getting ready to take off as soon as the airport's safety officer is done checking that the aircraft is safe to fly". Or "we will just be a few minutes while the Joint Safety Committee discusses our flight path".

The point being that safe work procedures need to be taken out of the safety department's hands and given back to the trades and professions who have the knowledge and education to develop them. It is this group who can also identify challenges with the procedures proactively and adjust to fit the real-time hazards on the job.

The Activity Trap and Reactive Measurement

OSHA, in the United States, defines leading indicators[51] as:

> *Leading indicators are proactive and preventive measures that can shed light about the effectiveness of safety and health activities and reveal potential problems in a safety and health program.*

In the OSHA guidance document, referenced above, OSHA describes four steps for using leading indicators. PDCA. They define this use as:

- **Plan** – Choose a leading indicator and set a goal for that indicator.
- **Do** – Communicate with your workers about the indicator, the goal and how you will track it.
- **Check** – Periodically reassess your goal and Indicator and
- **Act** – Respond to what you learn.

The concerning part of this approach, promoted by a regulator, is the immediate focus on goals (desired outcomes) as illustrated in Figure 4.1 above. In essence, they are recommending that leading indicators be approached in the same manner as lagging indicators. There also seems to be some confusion as to what is leading and what is lagging. We will explore this later in the chapter, but for now, it is worth examining how OSHA sees the PDCA approach playing out in a proactive manner.

Plan

In choosing the leading indicator, as listed in the plan, they suggest your choice of leading indicators should be based on data you already have (lagging), for example, the percentage of workers who attended the training. No mention of how effective the training was just how many complied.

Also included in the Plan stage is the setting of goals for this indicator. The example they use, related to the lagging indicator of how many workers did not comply with training requirements, is to establish a target of 100% compliance. The opposite of zero but the intent is the same.

Do

Here, they suggest collecting data such as training attendance sign-in sheets. Focused on the desired outcome of 100% attendance, there is no evaluation of the training itself or if it is successful or fit for purpose. Only compliance after the fact.

Check

In a proactive measurement, this would be established based on current situations. "Workers can't attend due to scheduling, workers don't attend because the training is irrelevant, workers attend but don't participate, etc.". These items are observable in real time. OSHA, on the other hand, remains focused on the desired outcome of 100% attendance based on lagging attendance records.

Act

The final step in OSHA's approach to leading indicator development and implementation is the comfort zone of react or reward. Focused on the desired outcome of 100% attendance, we forget to assess the real picture based on operational facts and miss the opportunity to improve on the delivery of training.

If this seems like a recurring theme as we move through the chapters that is because it is. Utilising lagging indicators to establish leading indicators is corrective action, not proactive management.

So how can we break this quality control activity trap and utilise leading indicators in the spirit of Heinrich's three fundamentals?

The first place to start is to ensure we know and agree on what constitutes a leading indicator. For the sake of this discussion, and simplicity, I will define each as:

Lagging Indicator = Reporting yesterday's news

Leading Indicator = Reporting current news that could impact tomorrow

The first being reactive while the second, if measured correctly, proactive. Both indicators have a distinct purpose, yet as discussed previously, we often time utilise one to confirm our assumptions about the other.

There is also a great deal of confusion as to what measurements constitute leading. When we start with our desired outcome as the goal, as suggested by OSHA et al., and zero harm is the go-to target, there is a tendency to dismiss nonrecordables, those things that are not part of the regulatory compliance paper chase.

Likewise, when we frame our goal as "zero harm" we conclude that nonrecordable such as first aid and near misses belong in the lagging indicator category as there was a slight injury or perhaps property damage. In theory, both those things are lagging in that the facts are gathered after the fact. Where they become leading is in the way we collect and respond to the data.

In my practice, I maintain that events such as near misses, first aids and property damage are in fact leading indicators. I almost always receive pushback on this belief from clients and fellow safety practitioners. The argument always goes something like this: "First aid cases involve injury . . . near misses are incidents . . . property damage includes loss . . . etc.". All true if you are looking through the lens of goal zero focused on desired outcomes rather than operational realities.

I had an HSE Manager for one client who insisted that they count first aid and near misses as events and track them in the same way they tracked TRIR etc. He maintained that they represented incidents and, no matter how slight, involved loss or harm.

I asked him how his first aid frequencies or rates were looking and he proudly said, "close to zero". This exchange was during a safety leadership programme I was facilitating at their company's head office in the US, which was attended by a hand full of managers from the head office and a hand full of field supervisors. So, I asked the field crew, "If I came onto your location today and opened a first kit what would I find?" The almost unanimous answer was something to the effect of "Not much".

People were having slight incidents all the time, but the consequence of reporting, perhaps the loss of a safety bonus, drove the reporting underground and created an environment of secrets. As the captain, and my boss, on the first drillship I worked on directed me on day 1, "Don't tell the bastards anything".

The reality is that counting minor nonrecordable with a goal of zero will inevitably turn those nonrecordables into nonreportable and, where leading indicators are concerned, the value is in the reporting driven by the positive consequence of doing so.

When we talk about events such as first aid being leading indicators, we are not talking about tracking them via a TRIR-type calculation as some compliance schemes require. We are instead focusing on them in real time based on facts not averages. The benefit here is that it allows us to see conditions that lead up to the event and correct them based on operations. They keep us focused on real-time opportunities not compliance based on a regulator's targets and goals or a client's contractor management software.

To do this successfully, we have to create a positive atmosphere around reporting. Make it easy, make it relevant, make it positive and make it matter.

To overcome the pushback often heard, such as that of the HSE manager mentioned earlier, is to recognise that not all leading indicators come from the same place. While, by definition, they are all preventative they are not all predictive. Some, like first aid, are corrective. What makes this last group leading versus lagging is the real time right now warning that something may be off, or work is drifting into the next zone. This is real-time information based on fact as opposed to manipulated calculations looking for averages and statistics based on what the watchman wants to tell us.

According to Andy Schneider, the global manager for safety at Caterpillar Safety Services, as referenced by Sandy Smith[52]: . . . *traditional metrics can help companies tell the score at the end of the game, but they don't help employers understand the strengths and weaknesses of their safety efforts and cannot help managers predict future success.*

In this analogy, it is a little late after the game to realise that when it is fourth down and you are 30 yards from the goal you should have kicked when you passed. That said, there is no guarantee that you will be successful if you had kicked. The blind spot here is that leading indicators are not about predicting success, they are about identifying opportunities based on operational facts, instead of comfort zones.

The only true prediction is . . . if your leading indicators are focused on a zero-injury goal your success will come in the form of non reporting and luck . . . until your luck runs out.

What we often see when safe by accident has run its course is organisations, rather than admitting the focus was wrong, will take their desired outcomes and apply them to a smaller more manageable aspect while maintaining the focus on zero.

So, if utilising lagging indicators to develop leading indicators creates blind spots how does utilising corrective leading indicators differ? The difference being that corrective leading indicators are derived from events that have just happened not desired results. They are front and centre on the plant floor and not being counted in an office or boardroom. They offer workers opportunities to implement corrective actions on the spot and capture lessons learned and extend the safe zone of work.

The challenge then becomes choosing the leading indicators that matter and prioritising them. This requires adopting new techniques and abandoning bad habits and comfort zones.

The key to developing successful leading indicators is starting with facts based on current operations and leaving the scorecards and compliance checklists to the safety department and regulators to worry about.

In Chapter 1, we looked at replacing the PDCA cycle with something linear such as ASPECT or CARE. This alone supports Heinrich's fundamental number 1 "The creation and maintenance of active interest in safety".

In Chapter 2, we looked at replacing the focus of management systems from rules to risk. Primarily, the concept of aligning tasks with the environment or zone they are conducted in.

Chapter 3 we touched on the triangle or pyramids meant for calculating cost and the challenges of implementing them out of context.

For the remainder of this chapter, we will return to the triangles and discussions focusing on operational realities that present precursors for serious injury and fatality, along with the need to filter out the high-frequency low-impact tasks that live in the safe zone (Figure 2.6).

Choosing Your Battles

As mentioned earlier, the day the crew of the Deepwater Horizon experienced catastrophic systems errors, taking the lives of 11 crew members, senior management from both the drilling contractor and the client were onboard celebrating seven years of nonreporting. This celebration of the safe zone drew attention away from the professional consultations being discussed regarding the planned operations which had begun drifting from the hazard control zone into loss control territory.

It is easy to sit here now in hindsight and see the flaw in management's focus on low-impact high-frequency events, like Lost Time Incident (LTI) rates and first aid frequencies, but at the time, it was commonplace to focus on desired outcomes – regulators, insurers and shareholders liked the statistical results it produced. I say it is easy to sit here in hindsight but what is not easy is to sit here writing this knowing that nothing has changed.

Despite millions of dollars' worth of regulator-funded research, we continue to focus on the bottom of 90-year-old triangles expecting different results, or worse, explaining away lessons learned as human error, bad culture or failure to follow rules.

This is especially frustrating when the regulator has reportedly estimated that more than 50% of serious incidents go unreported. And while they keep organisations focused on lagging indicators, they admit: *Too often, we would investigate a fatal injury only to find a history of serious injuries at the same workplace*[53].

Researchers are reporting that many industries and jurisdictions are reporting a drop in high-frequency low-impact incidents. This is likely the result of nonreporting. While it is easy to hide worker injuries and disguise lost time, it is not so easy to sweep fatalities under the rug. That is not to say that it does not happen; however, fatality reporting is likely more accurate than injury reporting.

This brings us back to Heinrich's triangle. In Chapter 3, we looked at two schools of thought regarding any of the safety triangles, pyramids and icebergs: the first being that given the ratios it is just a matter of time; the other being the belief that if we take care of the small stuff the big stuff takes care of itself.

The problem with the second school of thought is if we are strictly focused on the small stuff, we miss the leading indicators that point us to zone drift. In theory, we are focused on lagging indicator data to develop leading indicators. What is worse is they are using lagging indicators derived from reports that represent less than half of the operational reality.

There is, however, research out there that can help in the development of leading indicators utilising elements of the triangles and pyramids. Many researchers and practitioners have gone beyond the unsafe acts and conditions and the number ratios identifying that not all unsafe acts and conditions are created equal.

In 2011, BST and Mercer ORC[54] worked with seven multinational organisations to test the theory that reducing injuries at the bottom of the pyramids would in turn reduce these at the top. *(Note: BST is now part of the DEKRA Insight Group and copyright holder of the research finding)* What the study confirmed is what opponents of Heinrich and Bird et al had been claiming for years. Event causations cannot be correlated with outcome severity. You will remember the findings of Crystal Eastman from 1910 mentioned in Chapter 2 who stated:

> *It is impossible to state the total number of injuries during that quarter, because there is no available record except of cases received at the hospitals. But even were an accurate estimate of the number of injuries in a year possible, it would be of little value. A scratched finger and a lost leg can not be added together if you look for a useful truth in the sum.*

But the BST – Mercer ORC study went beyond the usual triangle dismissal and examined why focusing on the bottom of the triangle was flawed. They confirmed that safety management efforts designed to prevent the low-impact high-frequency events were not effective in reducing serious injury or fatality and that the majority of incidents at the bottom of the triangle did not have the potential to create serious injury or death. Not only did this mean that reducing incidents at the bottom of the triangle would not result in a reduction at the top, but it also further demonstrates the flaw in risk ranking based on frequency and the likelihood of consequence as discussed in Chapter 2.

They reported that incidents can be categorised into those with low potential and those with high potential for serious injury or fatality outcomes. The study indicated that only about 20% of incidents fell within the high-potential category. What this means is that the triangle has validity if understood and utilised in a proactive manner.

These events or activities in relation to zone management theory reside in the hazard control zone and above and can be identified based on work as imagined (within the hazard control zone) and work as done (as work creeps into the loss control zone). Identifying these precursors is a vital first step in developing leading indicators founded in fact.

In the chapters ahead, we will introduce conceptual models that can be utilised to align precursor identification to operational realities in order to facilitate the creation of solid leading indicator strategies for serious injury and fatality prevention aligned with zone management theory.

Bad Habit 5

Behaviour-Based Safety: The Problem Child

The concept of behavioural-based safety has been around since the 1970s. While there have been many references to the practice, and a great deal of research on behaviour as it relates to accident causation, many safety historians credit the term behavioural-based safety (BBS) to Dan Peterson, a researcher in organisational behaviour and management, whose work was, in part, based on that of B.F. Skinner, the father of behavioural analysis.

Others claim that the term was coined by E. Scott Geller, a behavioural psychologist, professor and researcher in the Department of Psychology at Virginia Tech College of Science in the United States.

Regardless of the origins or who tagged the process with the term, there has been no safety initiative in history that has been more misunderstood and implemented out of context.

The United Steelworkers of America[55], for example, have rightfully taken a very firm stance against what they call "Blame the Worker" safety programme and describe the use of BBS as follows:

> The term behavior-based safety is used to describe a variety of programs that focus on worker behavior as the cause for almost all workplace accidents. Simply stated, behavior-based safety proponents believe that between 80% to almost 100% of accidents are caused by unsafe acts.

Perhaps part of the misunderstanding is the interpretation of statistics by some safety practitioners that indicate human activity as a causative factor, confusing that with human activity as the root cause. It can also be attributed to the way in which many organisations packaged and sold behavioural observation schemes calling them BBS. Whatever the reason the United Steelworkers view, based on their consequence history, is not wrong.

In the late 1990s, as I transitioned from being an offshore medic/H$_2$S Technician to onboard safety officer at Cliffs Drilling, I did not have much in the way of tools to utilise in my daily routine of wandering the rig telling people to put on their safety glasses and hard hats, and then, one day a package arrived on the helicopter. In it was a box full of VHS tapes and stacks of workbooks and behaviour observation cards. With the promise of decreasing the company's TRIR and other lagging indicators and at the insistence of many clients, Cliffs had decided to implement a "BBS" system. To

DOI: 10.1201/9781003404958-5

achieve this as quickly as possible, they purchased a system called STOP (Safety Training Observation Program), which was developed and marketed by DuPont.

The VHS tapes in the package included a train-the-trainer series that I was to complete and five training tapes that I was to use to train the crew. The first step was to train supervisors to observe work and provide feedback (positive and corrective). The second was designed to teach workers how to identify hazards and personal behaviours, so they could self-audit themselves and their peers and the third step was to incorporate these newfound skills daily to watch out for one another and correct worker behaviour. To be clear, this is not behavioural-based safety.

The system also included observation cards that could be used to capture improvement opportunities and track the observations to create yet another leading indicator based on lagging information. The truth of the matter is STOP, and most observation schemes like it are just that observation schemes. While behaviour-based safety relies, in part, on observing work, worker observations are of little value if the root of the behaviours is not identified. The difference between observation schemes and BBS is that observations focus on correcting worker behaviour while a proper BBS system focuses on correcting the impacts of organisational behaviours.

Observation programmes fail for many reasons, and it is no coincidence that these reasons can be traced back to those that cause the systems failures we have explored so far in this book, which include implementing processes out of context, misunderstanding the intent, and managing the system as a process focused on desired outcomes rewarding, or reacting to the results.

In my career, since the time of my introduction to STOP, I have witnessed and been part of some of the most destructive practices applied with good intentions in the name of behaviour observation programmes and nothing can fill the PDCA activity trap like observation cards filled with faulty information. Sheets and sheets of charts and graphs representing conclusions based on information gathered from "the village watchman, who just puts down what he damn well pleases".

Let me offer a few examples. When we introduce any new system to the activity trap, there are always gaps that require filling and to satisfy this, we must always go back to the comfort zone of what has worked or, more accurately, what has been done in the past.

Applied to worker observation programmes, we *Plan* to capture and correct unsafe behaviours at the worksite. We *Do* this by having workers watch each other and filling out a card. We *Check* the cards for compliance, and track trends in the office, and we *Act* by rewarding compliance (quota) or react by restating the expectation (more quota). The gaps here are created by not having a well-defined criterion. Exactly, what constitutes an unsafe behaviour or condition that needs to be captured on a card to be counted by committees. What does compliance look like and what reaction or rewards are required to get it.

I once consulted on an offshore rig off the coast of Mumbai. The drilling company was an American multinational with drilling operations around the world. The plan they had was to create quotas of one observation per day per worker and to incentivise people to go over and above they offer cash rewards weekly for the person who completed the most cards. This approach was very successful in changing behaviours. Crew members who could not read or write began taking stacks of observations cards home with them on their days off to have their children complete them and then return to the rig for their 2-week hitch adding them to the stacks of cards filled out in the coffee shack or breakfast table in the hopes of winning the cash.

This programme failed as a fact-finding tool producing any sort of corrective actions based on fact, but the result was counter to creating any kind of active interest in safety it disengaged workers creating a reactive work environment based on "have to" efforts. This is what most organisations and safety consultants would refer to as a bad "safety culture". Another slick catchphrase designed to blame the worker.

It was commonplace on most rigs I served on to have one or two a day quota and to see workers filling out their daily cards at breakfast before putting on the work boots for the day. On one rig, I would be told to count the cards prior to the morning conference call with Houston, and if the count did not match the personnel onboard manifest, I was made to wake up those missing cards and make them fill one out. Talk about unsafe behaviour!

This gets even more out of focus when you add compliance issues. Many contractor management schemes require those seeking certification to implement an observation programme, and thereby remain a vendor of choice, to the larger client organisations. These certification or qualification schemes are focused purely on quantity with no follow-up or verification as to quality.

For every person walking around a job site observing work and checking boxes on a scorecard, there are at least three or four workers observing them and waiting to see how this process helps to fix the challenges they have in front of them daily. Those observing the scorekeeper have their own system of scoring and it does not include checkboxes.

Many behavioural science researchers subscribe to the belief that behaviours represent values bubbling to the surface. Some say it is the component of culture that we can see. So, what does it say about the subculture or professional culture of the safety profession when we allow and support organisations who sell costly systems that, by design, do nothing more than gather faulty information, focused on the desired results of regulators, while the workers fall into a state of moral disengagement? We are not promoting or supporting safe behaviours. We are demonstrating at-risk behaviours, which inevitably cause the work in the hazard control zone to drift into the loss control zone.

We do worker observations and check the inputs for compliance and trends, and we then act by applying antecedents in the form of safety campaigns and toolbox topics and/or we apply consequences in the form of rewards for compliance or punishment (usually in the form of quotas or loss of safety bonuses). We then sit back and either question why our process is not achieving the planned result, or worse, we celebrate our success based on lagging indicators forgetting that this entire cycle sends a clear message that it is advantageous to the worker to meet the quota and "tell them what they want to hear".

It is this type of organisational behaviour that BBS is meant to address. The value is in recognising when the systems are creating drift between work zones and identify systems elements that do not support an active interest in safety and provide meaningful corrective actions based on real-time needs.

At the worker level, these issues are driven by the many competing values such as the organisation's stated values versus their demonstrated values. It is the demonstrated values that define the organisational climate and create the social environment of the workplace.

At the organisational level, we focus on the issues as being a product of worker behaviour and revert to past processes we are comfortable with (safety campaigns, posters, rewards, etc.), all the while keeping to our comfort zone of Plan–Do–Check–Act. Because that is how the regulators tell us we need to do it. Regulators who graduated from the same schools as us attend the same conferences, hear what we hear and read the same research, yet lack the ability to take a chance and step out of their comfort zone.

At the end of the day, it is these comfort zones that create the research–implementation gap that is widening as our work environment and social constructs evolve with technology and a changing world. These are the very behaviours that BBS is meant to address not PPE compliance, housekeeping, trip hazards, etc.

In Figure 5.1, we demonstrate the activity trap in relation to change. Change means organisational change. Organisation refers to industry, professions, regulators and the like. In this instance, we are not talking about worker behaviour change we are looking at organisational leadership change. When leaders demonstrate change, workers follow. It is important here to remember that organisational change can be both positive and negative and the response by the workers will correlate.

Part of the reason organisations stay stuck in their process-driven comfort zones is that processes are busy work. We feel like we are accomplishing things. We can codify them into policies and procedures, and they give us tangible results, which we can measure by trending lagging indicators dressed up as leading. Yet values are also influencing results and they too can be measured by looking at behaviour.

An organisation's values are not those located in policy or values statements they are in the observable behaviours demonstrated by leadership in relation to the expectations placed on the workers.

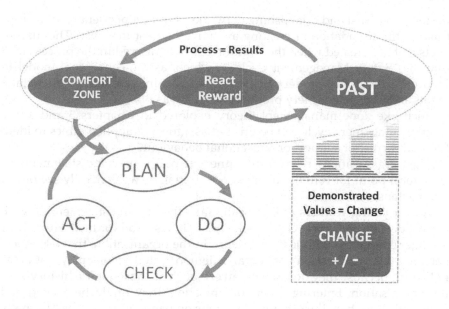

FIGURE 5.1
Process Versus Values

Diagnosing Organisational Behaviours

When taken to the workforce, the first thing a values or policy statement does is create awareness. Even if workers do not unilaterally believe that the statement will be backed by organisational support, they are, at the very least, made aware that there is a live choice to be made. For example, Do I report this hazardous condition? Do I stop unsafe work even though this means production will have to stop? The choice is a live one because there is some sense that the organisation will back the worker should they make efforts towards fulfilling the statement.

However, when organisations are focused on the Plan–Do–Check–Act-driven processes and are stuck in the "activity trap" and other comfort zones from the past, the worker has very little, if any, input into the "Plan" or the development of the policy or values statement. In most cases, the stated values of the company, or of those involved in the planning, are not aligned with that of the workforce. In this case, the statements become nothing more than an attempt to shape behaviours with a well-meaning but disconnected antecedent because the workforce has not been actively engaged in developing the policies and procedures that stem from it.

In other words, it is not something they are personally invested in. Yet it is precisely this sort of involvement in all stages of development that is essential for fostering trust and establishing networks of interdependent relationships that are the foundation of creating an active interest in safety. The disconnects that are created when the workforce is not engaged in the process feed into the "Safety Management Cycle"[56], which, as the conceptual model in Figure 5.2 illustrates, undermines even the best-intentioned mission statements and associated safety programmes, such as BBS.

Much like zone management theory explored in Chapters 3 and 4, the safety management cycle can be utilised as a form of applied ethics to identify and correct or support organisational behaviour.

When organisations rely on disconnected policy or value statements as their focal point for rolling out any safety initiative, we generally see one of two responses.

New or young workers will typically take the mission statement on with a great deal of zest. "Yes! Good! I like this! This is good for me!" For new or younger employees with less experience in the organisation, the disconnect between the organisation's stated and demonstrated values may not be as noticeable because they do not have an established consequence history with the organisation. Entering a new organisation, they might be excited and enthused about how the organisation looks on paper and how they formally present themselves and pride themselves on their commitment to safety.

On the other hand, more experienced workers, who have been with the organisational for a longer period, might be more sceptical: "Okay. I've heard this before. Show me you're actually going to back it up this time".

Either way, this rollout creates a moral awareness, which in turn prompts judgments within the workforce regarding the validity of the mission statement and the organisation's integrity in backing it up. These judgments are largely informed by their individual consequence histories. In many cases, more experienced workers will have seen this before. They will have seen mission statements and flavour-of-the-day schemes rolled out at different jobs, in a number of different ways, under a number of different names. Yet despite past discretionary efforts on their part and despite the stated goals of the organisation, they have not been supported, nor have they seen results, or the results have been negative or inconsistent with the alleged mission statement. This is their experience as they observed the observers, and it is a negative one. The judgments they form are largely derivative of these consequences histories, which in turn influence their individual forms and levels of motivation.

These variances in moral motivation are often exhibited through one of two responses. For example, new workers, with little organisational history to base their judgments on, are likely to be motivated by the antecedent. They believe in the values statement. This is good for them. They believe the statement will be supported by the organisation. On the other hand, the more experienced workers, those with a negative consequence history, will remain sceptical. These workers are most likely to be motivated because they "have to" do it. They need to comply with policies and procedures. In essence, their motivation is to avoid being punished. In either case, it is this motivation that drives the moral action and largely guides the form these actions take.

Following the rollout of the programme, those who are sceptical are watching the organisation's behaviour. They may be thinking, "Okay, let's try this. Let's see what happens". Those that have already bought in and want to do it take action and begin to implement the processes with vigour and enthusiasm. Eventually, however, competing values start bubbling to the surface, and these new workers start to realise, "Wait a second, I'm getting conflicting stories about time, money, and production" or "I'm putting in all these observations, but nothing's really changing for me. The company is saying "We're going to be world-class in safety", but I've got all this other stuff that seems to be taking priority". They are faced with competing values that are often difficult to recognise in a process-driven environment, creating dilemmas that do not get addressed or resolved and their work is drifting away from hazard control zones. In these sorts of climates, motivated workers seldom receive positive feedback or support from the organisation or their less motivated peers. Over time, this establishes a negative consequence history for the less experienced workforce, which is not easily undone unless directly addressed. Thus, if left unattended, regardless of what the new worker's moral judgments and motivation looked like when the programme was rolled out and the values statement originally issued, an erosion of trust in the integrity of the organisation begins to occur.

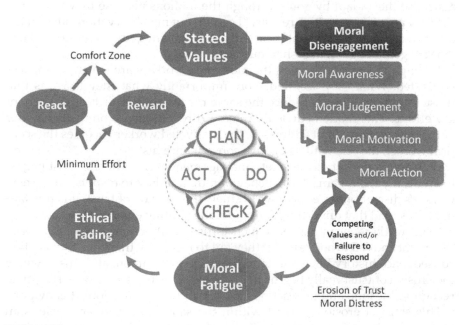

FIGURE 5.2
The Safety Management Cycle

This erosion of trust, however, is not solely the product of the direct responses the individual receives from his or her leaders. There are other factors influencing the novice worker's trust in the organisation beyond their direct relationship with its leaders. As part of an organisation, the individual worker is embedded within a larger network of relationships that constitute the situational context. Within this situational context, there is a multitude of actors whose behaviours and attitudes constitute, teach, reveal and convey the hidden curriculum of the organisation.

This hidden curriculum is a composite of the informal rules and organisational norms that implicitly inform members of "the way things are done around here". It forms the social constructs that define the organisational climate.

Think back to those young workers who bought into the vision statement straight away. They strive to implement the new programme wholeheartedly. They start out motivated by discretionary efforts, and they want to see positive results and change. However, it is often the case that any rewards for these discretionary efforts are shared with those who demonstrate little or no effort. This shows the novice that the consequences are inconsistent with the values statement; that is, "You tell me that we're going to be world class in safety, but the messages I'm seeing don't match the antecedents that you've got up on the wall".

The sceptics, on the other hand, approach it half-heartedly demonstrating have-to efforts. Their consequence history tells them this is just another false alarm, so they adapt by going through the motions because they "have to". Yet they are seeing positive results. "I can fill out my observation cards in the lunchroom. I can do this half-heartedly and I'm still going to get that jacket. I'm still going to get that cash bonus".

Through this entire process, the motivated workers are watching the less motivated. What they see, and more importantly what they learn, is that those minimum efforts produce the same results and rewards as discretionary efforts. This hidden curriculum has a direct influence on the behaviour of the new worker. If not addressed, the motivated worker becomes the sceptical worker, the one with a negative consequence history.

Within this pocket of the safety management cycle, wherein trust begins to erode, it is the organisation's behaviour, their failure to respond to the new worker's discretionary efforts, which fuels this erosion of trust. It is this failure to respond that unwittingly supports the tainting effect of the more experienced workers' negative consequence history. When organisations get to this point and fail to address it, if they fail to recognise this loop, the sceptics' consequence history begins to taint that of the champion, and this is where the erosion of trust really begins to take hold. This is also the opportunity to recognise the loop and take actions to correct the organisational behaviours.

This stage of erosion of trust within the safety management cycle is an organisation's first opportunity to effect change. It is the starting point to begin asking the hard questions, to get out of their comfort zones, and identify the values they may be unknowingly demonstrating which are contributing to this tainting effect. It is a leading indicator which helps to identify when leadership may not be responding in ways that are consistent with their stated values. In effect, this cycle can be tied back to the values statement. We know when we come back to reissue a statement and start this whole process again that this is going to be at play. We know the sceptics are out there. Before rallying the troops, we first need to deal with that infection.

This erosion of trust, if left unaddressed, leads to a form of moral distress, as workers must continually try to balance these seemingly incommensurable values. Their waning trust in the organisation further exacerbates their distress because at the same time they are likely to begin questioning how their efforts are going to be responded to, whether their discretionary efforts are something for which they will be rewarded or whether in fact their efforts are something for which they might be (informally) punished.

Over time, this recurring feeling of moral distress leads to what applied ethics terms moral fatigue. Morally fatigued individuals are often thinking,

> I want to do this. I truly want to do this, but no one's responding to my efforts. I'm seeing those around me being rewarded for just getting by, and I'm tired of exerting all this energy and dealing with the stress of it all.

Moral fatigue is the impact over time of feeling morally distressed, and it has both personal and organisational consequences.

If not corrected, moral fatigue moves into ethical fading – where the value of safety becomes neutralised. We no longer notice the ethical dimensions of acting safely because there are other competing factors – like money, time, production or even holding onto your job. So, the ethical dimensions begin to fade from view.

When the value of safety is neutralised, those workers who were keen in the beginning become sceptical workers, discretionary efforts morph into have-to efforts and values statements start being seen as flavour-of-the-day. Those workers who are sceptical to begin with were not always the sceptical workers. When they first started, when they were first introduced, they took it on as champions. It is this erosion of trust, and this cycle of moral distress, fatigue and neutralisation that creates those sceptics. Now we are in a position in which our antecedents, our values or policy statements, designed to move sceptics to discretionary efforts, are moving our champions to have to efforts. Just enough to get by.

In the absence of discretionary efforts, we are led to a situation in which only the minimum efforts are being exhibited. The workforce is complying because they have to. They no longer champion it as their own because the ethical fading has taken out or neutralised the value of safety. They are just not seeing the support. These minimum efforts either lead to a critical incident that compels us to react, or the organisation rewards outcomes based on reported lagging indicators, which may or may not be accurate. Either way, in the reacting or rewarding, we will often return to our comfort zones as we move forward, positioning us once again in the activity trap. In doing so, we continue to neglect our values, and how, if at all, we are demonstrating those values. We miss the opportunity to uncover how our competing values, for better or worse, are driving decision-making and end results.

These competing values are truly the elephant in the room. They are always there and always looming large. We say safety first, but we know full well that, money, time and production are always going to be factors. The workers know it. The organisations know it.

The activity trap of Plan–Do–Check–Act impedes organisations from dealing with this head on and thinking about what this looks like in practice and how they are going to address the competing values when they arise. Out of desperation, well-meaning safety practitioners return to their comfort zones of process-driven desired results-oriented safety management. The real issue remains the unaddressed elephant in the room, and organisations continue to apply more antecedents, all the while failing to deal with the consequence history that they have created through stating a value, and demonstrating something else.

Often, organisations will react to less than favourable outcomes by suggesting that workers need a refresher course in soft skills such as BBS, team building or safety leadership and the like. Following these soft skills training programmes, it is not uncommon to see workers leaving feeling invigorated and motivated. However, this is seldom, if ever, sustainable once they get back into the comfort zone work environment. This is due, in part, to the organisational response to minimum efforts by restating values they have failed to demonstrate in the past. This renewed verbal commitment in response to the effects of the safety management cycle, while intended to motivate discretionary efforts among the workforce, actually creates a morally disengaged workforce.

This moral disengagement is a direct result of ethical fading, and when workers disengage, they begin to turn off their usual ethical standards. This is to say that when workers are disengaged, they shut off their own personal codes of ethics within the working environment They are one person at home, in their communities, with their families, and another person at work. With different roles come different sets of morals (i.e., role morality).

In order to manage or mitigate the cognitive dissonance created by not acting in accordance with their own personal code of ethics, the worker may begin to rationalise that the risk of getting hurt on site actually is not all that great. It is the moral disengagement coupled with a neutralisation of the value of safety in the workplace that makes this splitting possible.

When organisations do these refresher courses or recommit to values statements, when they put those antecedents to work, moral awareness, judgment and motivation are all impeded because the workforce has disengaged. The effect is going to be even less than it was the first time. When organisations fail to back up antecedents with positive consequences, renewing organisational commitments can be even more dangerous than not doing so, if the negative consequence history is not first addressed and these statements are not actually put into action.

It is the observable behaviours in response to safety systems that we use to diagnose the effectiveness of our systems and our operational environments. This diagnosis does not include pulse surveys, cultural assessments, employee interviews or any other such "tell me what I want to hear" methods. It involves walking around watching, listening and learning and, when implemented correctly, supports Heinrich's three fundamentals of creating and maintaining active interest in safety, fact finding and corrective action based on the facts.

While managing risk with rules is flawed the reality is that, in most professions and trades, we must have procedures and guidelines built on lessons learned and best practices. Many of these revolve around creating a safe work environment, and while not the end all and be all to managing safety, there is a time and place for them. Speed limits, fall protection, safety eyewear, etc., all serve a purpose and, this has become commonplace in many industries and societies. The danger they present in relation to the safety management cycle is when they are not fit for purpose or constantly enforced. Nothing will push a motivated worker toward ethical fading faster than inconsistent leadership.

This is one area where observing worker behaviour, or more accurately, team behaviours can help to identify and thereby control the drift from moral distress to moral fatigue. This comes in the form of identifying behaviour issues in the field and determining the root cause.

Why workers fail to do what they are told or directed to do has been under the microscope since the start of the human era in safety. In that time, sociologists and psychologists have developed many theories, diagrams, models and opinions on what drives poor performance, but when examining work-as-done, the issue can be divided into three cause categories:

Errors – Erosions of Standards – Violations

In a rules-based safety environment, we tend to focus on violations and when we react or reward and apply the consequence, good or bad, to the entire team. As mentioned earlier, rewarding those who put in minimal effort with the same reward given to those who demonstrate discretionary effort and go above and beyond causes them, over time, to join the ranks of the unmotivated minimal effort workers. This is why it is important to understand the issue before reacting and rushing to the do.

Violations happen when a worker knowingly and wilfully violates a clear policy or rule. These are individual worker issues that need to be dealt with by the human resource department in a consistent manner in private. Far too often I have heard a safety officer on site brag about how they had to "rip someone a new one". Policing safety is ineffective, and discipline is the domain of HR and leadership, not safety.

Likewise, we know that people make mistakes and unintentionally fall out of step with policy or procedure. There are many reasons for this such as a lapse in concentration, not being aware of a policy, not being trained properly or simply not having the capacity to perform the task. Regardless of the reason, mistakes happen and we live in an imperfect world. As with violations, I cannot count the number of times I have seen a safety practitioner reprimand someone for making a mistake. Nothing will deflate a motivated team member faster than being reprimanded for a lapse in skill, knowledge or judgement.

The difference between a violation and an error is that violations are worker issues while errors are systemic leadership issues. Errors account for about 20% of all performance issues, while violations account for 5% or less. This leaves us with the majority of performance issues unaccounted for. These belong to the domain of eroded standards. (Figure 5.3)

An erosion of standards is much like a violation in that the worker knowingly breaks from policy or procedure or rule; however, what differentiates an erosion from a violation is that deviating from the rule has become commonplace and many workers do the same. Or the rule is unclear and not fit for purpose or competes with other rules, and most often, the erosion is the result of rules not being consistently enforced.

While violations are worker issues that must be addressed by the HR department, erosions are leadership issues and must be corrected by leadership. That is to say that it is the leadership or organisational behaviour that must change to correct the problem.

Leadership Issues		Worker Issues
Error 20% Unconscious Non-Compliance	Erosion of Standards 75%	Violation 5% Conscious Non-Compliance
Worker unknowingly breaks policy, procedure or rule OR Worker incapable of successfully executing the work OR Worker makes honest mistake (refer to Human Factors)	Worker knowingly breaks policy, procedure or rule, but: • Policy, procedure or rule is not clear, or conflicts with other rules • Breaking policy, procedure or rule is accepted site practice	Worker knowingly violates clear policy, procedure or rule

FIGURE 5.3
Performance Issue Causation

Within any organisation, erosions exist and while erosions are normally being seen in the behaviours of more than one worker not all team members take part in the shortcuts. There are always those who are still motivated to do the right thing because it is the right thing to do. The danger is that erosions are contagious and if left unchecked will either infect your motivated team members to the point of moral disengagement or the moral distress and the subsequent fatigue they feel witnessing their teammates step outside the rules while leadership does nothing forces them to find employment elsewhere and your best players leave to join another team.

In the next chapter, we will explore how redefining joint health and safety committees can address and monitor erosions of standards by diagnosing and monitoring organisational behaviour and building on the concepts from this and previous chapters creates a living system to support the three fundamentals of accident prevention.

Bad Habit 6

Joint Health and Safety Committees:
Put Down the Donuts There's Work to Do

Globally joint health and safety committees are known under a variety of names depending on the jurisdiction which requires them or the organisations that wish to develop them. For our purposes, we will refer to them as Joint Health and Safety Committee (JHSC) or at times simply the committee.

The Canadian Centre for Occupational Health and Safety (CCOHS) defines a JHSC as "a forum for bringing the internal responsibility system into practice". The internal responsibility system being a system that puts in place an employer–worker partnership in ensuring a safe and disease-free workplace.

The Occupational Safety and Health Administration (OSHA) in the United States describes a safety committee as a committee comprised of employees that are concerned about potential and real hazards in the workplace and encourages co-workers to follow safety protocols and recommends best practices to mid- and upper-level management.

The International Labour Organization (ILO) defines them as a bipartite body composed of workers' and employer's representatives, which is established at the workplace and is assigned to various functions intended to ensure cooperation between the employer and workers to achieve and maintain safe and healthy working conditions and environment.

Regardless of how they are defined the concept of enacting a joint health and safety committee in any organisation is one that, in theory, can form the backbone of demonstrating and advancing values to produce a climate of trust conducive to creating sustainable step-change in both business outcomes and safety performance. It can and should be the foundation of Heinrich's first fundamental of accident prevention, creating an active interest in workplace health and safety.

The reality however is quite different in most organisations. While the intent is to create and maintain an active interest in safety, thereby creating an opportunity for leadership to demonstrate organisational values and support discretionary "want to" efforts, many committees unwilling and unknowingly create the opposite. Committees formed for committee sake have a tendency to hide behind closed doors and, while they may well be doing busy work, the workers fail to see responses to their concerns and leaders fail to see a return on investment leading to moral fatigue, ethical fading and minimum efforts all around.

  DOI: 10.1201/9781003404958-6

To better understand this apparent disconnect, we need to first examine how they are often implemented out of context. That is to say that while committees, in theory, are meant to support discretionary effort, their very creation is the result of "have to" effort. Most jurisdictions require organisations of a certain size, under law, to develop a joint health and safety committee. In many jurisdictions, this requirement can also be imposed on smaller organisations as a reaction to high worker's compensation claims or reported safety violations. Either way, the committee in and of itself is often developed as a reaction to lagging indicators and as such becomes focused on desired outcomes rather than operational realities, thereby feeding the safety management cycle discussed in Chapter 5.

As Dr Dominic Cooper wrote in his book, Improving Safety Culture – A Practical Guide[57]:

> *Most people are not aware of the issues being discussed by committees, or what the results of their deliberations might be. As a result, although those who attend safety committee meetings may well be working hard, safety committees tend to be seen as coffee drinking forums rather than locations for serious discussion about prevailing safety issues.*

In the book, "Value-Based Safety Process – Improving Your Safety Culture With Behavior-Based Safety"[34] referenced in Chapter 2, Dr Terry McSween says it best:

> *committees have difficulty in developing firm plans for achieving safety improvement . . . Rather they continue to plan training for toolbox safety meetings and revise safety awards while implementing yet a further variety of programs that rely on the use of posters, slogans, and other approaches of dubious effectiveness.*

So how can organisations develop firm plans, based on operational realities that will yield a return on this "cost of doing business"?

Most jurisdictions that have legislated requirements for such committees also have requirements outlining how they must operate. Committee responsibilities differ somewhat between jurisdictions, however, most revolve around compliance and desired outcomes and draw the committee's attention away from operational realities. It is not the intent of the committee to replace safety managers, and it is not the place of the safety department to run the committee but more often than not the joint safety committee becomes seen as the "go to" platform to get the attention of leadership.

Committee Responsibilities – Blind Spots and Activity Traps

A good first step in developing a committee's terms of reference, or playbook, is to look at the legislated roles and responsibilities and define them. This is an area where a safety professional can and should assist in the role of an advisor to the committee not as a committee member or chairperson. While the committee is not the safety professional's domain having a third party who can help navigate safety legislation from an advisory position can be very helpful in satisfying the legal requirements while remaining focused on the operational realities. It will allow the committee to delegate the compliance aspects to the safety manager and allow the committee to focus on real-time issues.

Let's be honest, the first safety practitioner who can convince a regulator they need to change will be a worthy recipient of a Nobel Prize. When committees are seen to be focused on compliance it adds to the moral fatigue caused by a failure to respond to the real challenges faced on the job. Utilising the safety manager or human resource manager to run interference on compliance is a good use of their resources in support of the committee.

To demonstrate this in practice let's look at some of the more common roles and responsibilities listed by regulators in the UK, United States, Canada and Australia and identify the blind spots these create. While this collection of 14 common responsibilities is not inclusive of any one jurisdiction, it will help to demonstrate how we can satisfy the requirements while not being seen as the "safety manager" committee.

Once we take a deep dive into these challenges, we will talk about developing strategic proactive plans utilising information gained from human factors research and the concepts discussed in the previous chapters. This will allow the reader to facilitate the development of firm plans for achieving safety improvement beyond slogans, posters and other comfort zone activities leading to dubious results. This is not easy work. It requires all sides, management, workers, union reps and sometimes contractor personnel, to take the difficult first step outside their comfort zone and past bad habits. It also requires creativity on the part of the safety department to run interference on compliance-focused third parties.

The one common thread these duties and functions have is that they take the responsibility off both the employer and regulator to ensure the organisation is compliant and creates a wedge between the workforce and committee members. Like most things "safety" we have somehow managed to drive every well-founded idea into an activity trap of legislation and rules-based compliance and busy work.

Perhaps it is the hangover from Taylorism or the belief that the only actions that matter are those that can be quantified and entered into a database at a government office. Moving regulators off their comfort zones is never easy and most time downright impossible. This is not political, it is bureaucratic and the only way to change it is to be the first to explore outside the comfort zone. This takes bravery and must be done in a professional and respectful manner, but at the end of the day, as in all bureaucracies, forgiveness is easier to get than permission.

TABLE 5.1

14 Common Responsibilities Required by Regulators

	Required Activity	Trap
1	Identify situations that may be unhealthy or unsafe for workers and advise on effective systems for responding to those situations.	Committees get focused on unsafe acts and conditions related to high-frequency low-impact events driven by desired outcomes and other matters of compliance.
2	Consider and expeditiously deal with complaints relating to the occupational health and safety of workers.	This creates a dependent work environment where committee members are seen to be responsible for safety.
3	Consult with workers and the employer on issues related to occupational health and safety.	The committee can often become disengaged from both the workforce and management and viewed as an entity outside of operations.
4	Make recommendations to the employer and the workers for the improvement of the occupational health and safety of workers and compliance with the occupational health and safety regulations and monitor the recommendations' effectiveness.	This drives the committee to a role of control and creates a division between the committee members and the workforce. Committee members become seen as the "safety police".
5	Advise the employer on programmes and policies required under the regulation for this workplace and monitor their effectiveness.	This is and should be a function of the safety and/or human resources department.
6	Participate in inspections and inquiries as provided by the regulation.	Inspections become commonplace and condition the workforce to "clean up" prior to the regular inspection, usually following the committee meeting.
7	Make recommendations to the employer on educational programmes promoting the health and safety of workers and compliance with the regulation and monitor the recommendations' effectiveness.	Safety training is often limited to committee members. Developing competency progression and training matrix should be a human resource function and safety training should be redefined as trades or job training.
8	Advise the employer on proposed changes to the workplace or the work processes that may affect the health or safety of workers.	Operations meetings are most often more regular than safety committee meetings and the workers' right to participate is better demonstrated in relation to their work.
9	Ensure that incident investigations and regular inspections are carried out as required by the Regulation.	This silos the committee into a separate entity and again creates a dependent environment.
10	Select appropriate worker and employer representatives to participate in preliminary and full incident investigation processes.	Preliminary investigations should not be held up waiting for committee members while full investigations require specialised leadership. Committee member involvement should be limited to advocating for those involved in an incident.

(Continued)

TABLE 5.1 *(Continued)*

14 Common Responsibilities Required by Regulators

Required Activity	Trap
11 Review and provide feedback on any corrective action reports resulting from incident investigations.	Corrective action tracking should be monitored at operations meetings which are separate from committee meetings.
12 When necessary, request information from the employer about known or reasonably foreseeable health or safety hazards to which workers at the workplace are likely to be exposed.	It is the employer's responsibility to provide this information and the workers' right to know. Ensuring this is done should be a safety/human resource manager function.
13 Request information from the employer regarding the health and safety experience and work practices and standards in similar or other industries of which the employer has knowledge.	Applying this function to the safety committee removes the opportunity for the employer to demonstrate stated values at the worksite.
14 Carry out any other duties and functions prescribed by the Regulation.	Functional Terms of Reference require clear expectations and roles. "Other duties prescribed by regulation are ambiguous and often lead to the committee becoming the dumping ground for the activity trap do's"

What's in a Name

"If Moses was a committee the Israelites would still be in Egypt"
Source Unknown

As mentioned at the top of the chapter, joint health and safety committees go by many names depending on jurisdiction and I have seen many well-intentioned committees get stuck agreeing on who they are, what they do and how they do it.

The problem all committees face from the start is just that . . . they are committees. By definition, a committee is a group of people appointed for a specific function, typically consisting of members of a larger group. While a joint health and safety committee's function is to support the creation of a safe and healthy work environment, the varying nature of the roles and responsibilities dictates that the committee's specific function is to support desired outcomes rather than supporting and maintaining an active interest in safety among the members of the organisation. I have worked with committees who called me in simply because they had been stuck for years developing a simple term of reference. I personally advocate that the only terms of reference a functioning committee needs are Heinrich's three fundamentals of accident prevention.

1. The creation and maintenance of active interest in safety
2. Fact Finding
3. Corrective action based on the facts.

But again, a committee is a committee, and as such, legislators expect them to function in the same manner as political committees. Robert's Rule of Order, etc., becomes the focus, and as such, the committee begins to morph into a bureaucratic beast with representatives sitting on opposite sides of the table. Not at all conducive to creating an active interest in safety, fact finding or focusing on fact.

So how do we change this and if we change it what does the committee do if not dot I's and cross T's?

Step one in breaking free of the committee's comfort zone is to rename the group. Change the approach from being a Joint Safety Committee to being a Learning Team. Dropping the word "committee" entirely and replacing it with team instantly turns two or more sides into one. A team works together to get things done, shares a common purpose, shares best practices among themselves and has each other's backs.

According to Dr Todd Conklin[58],

> *A Learning Team is a facilitated means of engaging with workers to understand and then learn from the opportunities that are presented by everyday successful and safe work as well as learning from events or incidents. This includes understanding what, when, how and why people do things differently from the formal systems and procedures in order to get the job done.*

So how can we apply this to a proactive approach in creating and maintaining an active interest in safety focused on fact finding and corrective actions based on facts? In the book "The Practice of Learning Teams"[59], Brent Sutton, Glynis McCarthy and Brent Robinson outline five core principles of learning teams as follows:

1. Understanding that Work-As-Imagined and Work-As-Done give context.
2. Groups outperform individuals in problem identification and problem solving.
3. Workers have the best knowledge and understanding of the problem.
4. The more effort put into understanding the problem, the better the solution outcomes.
5. Group problem identification, solving and reflection (soak time) drive learning and improvement.

The above principles help to dictate the spirit in which the new team operates and helps to align fact finding with work as done. Aligning your learning teams (committee) with your operational reality is not as difficult as many believe, but it does require stepping out of the comfort zone and reframing questions to get an accurate picture of the operational realities.

Step 1   Create a structure which supports collaboration, transparency, respect, and trust.

Step 2   Identify work activities by zone and the elements that drive work out of the safe zone and the controls in place to manage the risks and zone drift.

Step 3   From the activities determine the related elements that represent precursors to serious injury and fatality starting with the activities in the out-of-control zone.

Note: here you are looking for the element not the actual job Eg. if painting from a ladder is the job, working at heights is the element.

Step 4   List the controls in place for the precursors and identify any gaps. For example, working at heights controlled by fall restraint or fall arrest.

Step 5   Identify those controls which have eroded. Your best source for gathering erosions is from those doing the job. Even those participating in the erosions don't necessarily like them and will willingly share their concerns if they are given the opportunity, a safe way to do so and the belief they will be acted upon.

Step 6   Align the list of erosions to the list of precursors and prioritise actions based on the zone where the activities in question reside.

Step 7   Coach leadership in methods to improve the operational climate in support of erosion management and post-normal safety collaboration while monitoring for erosions of trust and moral distress.

This process is not a magic bullet and will not happen overnight, but the initial benefit is your new team, previously the committee, has established leading indicator targets based on facts and challenges encountered at the jobsite.

Erosions can happen quickly if leaders are unaware, but they take time to correct. It is critical that the directives for correction come from leadership and not the safety department. These directives must be communicated to all workers at the same time so that the entire team hears the same message.

It is critical that the leader:

1. Be specific and base their concern on facts.

2. Communicate the reason why the rule or directive is in place.

3. Come from a place of caring not rules. People need to know you care before they care that you know.

4. Be clear with their expectations going forward and engage the workers in the solution.

5. Draw a clear but realistic line in the sand including consequences for those not willing to participate in the correction.

6. Be prepared and able to implement the stated consequences when the line is tested, and it will be.

7. Monitor the progress and celebrate successes.

Correcting erosions, one at a time, takes patience and could take six months or more. It is not uncommon for learning teams to be faced with a rather extensive list initially when undertaking the alignment activities and while six-plus months seems like a long time to focus on one issue what we find is that, once leadership has corrected two or three successfully the crew members start to correct the others as a matter of "the way we do things here".

One last blind spot to be mindful of during this process is to ensure the learning team (committee) does not get sidetracked and focused on compliance, lagging indicators, routine inspections and the like. These items are important but best left to the safety department. It can often be helpful to retain an outside resource to help guide the process in the early stages. Not lead the team but rather be that coach on the sidelines who can watch the plays and help keep the team focused.

Simply changing the way in which we frame the intent of our activities as a team begins the difficult journey to creating and maintaining active interest in safety, fact finding and implementing corrective action based on facts.

Taking on this difficult work will truly create and maintain an active interest in safety supporting fact finding based on work as done and facilitate proactive corrective actions based on facts. If managed correctly, it shapes the work environment to that of an interdependent and benevolent organisational climate, which we will now examine in our final chapter.

Bad Habit 7

Safety Culture: If You Can't Speak The Language Blame The Culture

No book on safety would be complete without a word or two on safety culture.

I have spent the better part of my career working in multicultural environments, usually as a guest in someone else's culture. I have also spent the last years of my working life in my country of birth working with some of the first cultures and I now live in a country with a strong Spanish cultural influence on the land of another first culture.

So when I hear the term culture used in relation to safety, I can honestly say that in all the cultures I've lived and worked in I have never encountered one that was safe or unsafe by definition. In this last chapter, we will take a quick look at the components of culture, and we will skim the surface of applied ethics as it relates to decision-making and therefore safe or at-risk behaviour.

My personal reflection on defining both safety, and culture, takes me back to 2009 when I researched and published my dissertation in partial fulfilment of my Master of Science degree at The University of Greenwich in the UK, a foreign culture to mine.

My research examined the relationships between national culture, personality traits, risk perceptions and at-risk behaviours. At that time, I defined "safety" as "the absence of harm". Not very original I know but it seemed to fit the narrative of the day. What I now find interesting looking back was the definitions I used to define "culture" and how it literally applies to that of safety.

To avoid confusion between the two concepts, I feel it important to first revisit early definitions from my 2009 work and then share how these have changed throughout my career observing these concepts being applied in the field.

Culture

Many researchers and business leaders, such as Jeanie Duck, believed that the "culture" from which a multinational organisation originates, "the home

DOI: 10.1201/9781003404958-7

culture", will dictate the culture of the organisation. For example, Chevron and apple pie versus Shell and wooden shoes. This belief was also shared by Geert Hofstede[60] in his 2001 publication *"Culture Consequence: Comparing values, behaviours, institutions and organizations across nations." (2nd edition)*

On the other side of the coin, there are those such as Jeffery Arnett[61] and Katherine Mearns and Stephen Yule[62] who share the belief that the national culture of the workforce and the geographical location they operate in are more likely to influence behaviour.

It is my belief, after 30-plus years of working and living on five continents, that both these findings are incorrect perhaps because of definitions. While one group was looking at culture, the other seems to have been focused on operational climate.

There have been many definitions assigned to the word culture. They range from the very inclusive Herskovits' 1972 definition[63]: *"The Human made part of the environment".*

To more focused definitions such as Shweder and LeVine's 1984 definition[64]: *"A shared learning system".*

Geert Hofstede, considered by many to be a pioneer in the study of culture, wrote in 2001 that culture can be defined in two ways. First, it is the way in which people's minds are trained or refined within a civilisation and, second, it presents a collective way of thinking, feeling and behaving. *"A collective programing of the mind".*[60]

Considering Hofstede's description, it is no wonder that we get caught up in posters, slogans and buzzwords when we hitch safety to the culture wagon, it does however explain why those bent on "creating a safety culture" fail to produce sustainable improvements in safety performance. Outside of the military and organised religion, I cannot think of an organisation I have been part of that successfully programmed the team members' values and beliefs. When we continue to try in the name of safety culture, we are doing little more than "safety colonisation".

Culture can be described as a noun: *"the integrated pattern of human knowledge, belief, and behavior that depends upon the capacity for learning and transmitting knowledge to succeeding generations"* Or *"the set of values, conventions, or social practices associated with a particular field, activity, or societal characteristic".* Or a verb: *"to grow in a prepared medium".*

Safety

The word "safety" is a noun. Nouns name people, places, ideas or things. So, when people say they "do" safety are they saying safety is a thing? How many times have you heard a supervisor or manager refer to safety as a person? "Better call safety in here to do a tailgate meeting". And of course, let's not forget safety as an idea "Safety First!"

A noun is defined as a word (other than a pronoun) used to identify any of a class of people, places, or things (common noun), or to name a particular one of these (proper noun). According to Webster's, "Safety" is classified as a common noun because it does not give the name of a specific thing.

Safety Culture

When safety practitioners and others use the term "safety culture", it is safe to say that, most often, they have not reflected on the meaning. This is common when speaking in catch phraseology and flavour-of-the-day campaigns. We cannot define what it is, but we spend millions looking for it and trying to measure it.

Before we take a deep dive into the literature surrounding culture and the many reasons safety is not cultural, let us assume that some practitioners view safety culture as a double noun: *"(safety) creating the condition of being protected from or unlikely to cause danger of risk or injury through (culture) a set of values, conventions, or social practices associated with a particular field, activity, or societal characteristic".*

While others, scrambling to "do" something, may approach it as a common noun and verb:

> *(safety) creating a state in which hazards and conditions leading to physical, psychological or material harm are controlled in order to preserve the health and well-being of individuals and the community by (culture) changing or growing values and practices in a set of prepared systems and directives applied to the worksite.*

Put into practice, regardless of the noun or verb application of "culture", the reality is that the organisation will be caught up in the activity trap and rushing to the comfort zone to "do" something. As demonstrated in Chapter 1, comfort zones rely on the past. Specifically, the organisation's or safety practitioner's consequence history. All the directives, initiatives, slogans, and soft skills training we make workers endure in the name of safety culture are nothing more than a collection of stories through words and images. By this definition, it appears that comfort zones better fit the definition of a professional subculture.

This is probably closer to the truth as we will explore later when talking about subcultures or subjective cultures such as those that are profession specific. How ironic would it be if the catchphrase utilised to justify systemic failures in safety is actually created by the subculture of the safety profession? This might account for our insistence on the continued use of the term safety culture and the millions of dollars being spent, year after year, to establish it as something that is measurable and malleable despite mounting evidence and failures to the contrary. It appears the problem is not with a safety culture but rather the safety profession. The seven bad habits do not describe worker activity and choices, they detail systemic failures propped up by the safety profession.

Well-meaning safety culture entrepreneurs and organisations unwillingly find themselves believing that all they need for success is for everyone to fall into line with their beliefs, traditions, symbols, values and rituals. "Forget what you may know or believe or have been told". Forget what you've been taught or successfully implemented and adopt the organisation's traditions, beliefs, and values. Follow the new leader's symbols and rituals. This is what I now refer to as "safety colonisation", and like national culture colonisation the world over, it never works and always has disastrous and deadly results.

Before we change direction completely and examine ethical and operational climates, I want to dive a little deeper into culture and then climate as a base to help us keep the two concepts separate.

To find common ground among culture researchers during my dissertation research in 2009, I landed on four characteristics that consistently appeared in the literature review.

1. Culture is a collection of individuals who share common values, beliefs, ideas, etc.
2. Culture is not individual, but individuals learn the culture of a group when they become a member of the group and it is transmitted from generation to generation.
3. Historical dimension is a related aspect of culture.
4. Culture has four distinct layers.

To keep focused on culture versus climate, I will use these four definitions as common ground to define culture when applying the term to safety. Also, I feel it important to define what culture, in its simplest form, looks like and will reference Geert Hofstede's[65] cultural layers model from 1991. The model is represented by four layers, values, rituals, heroes and symbols (Figure 7.1).

The limitation of applying this model is that many believe it represents each layer as being static. In 2006, the work conducted by Leidner and Kayworth[66] concluded that changes between various types of culture are dynamic.

Most researchers agree that culture can be divided into four distinct types.

1. National Culture
2. Organisational Culture

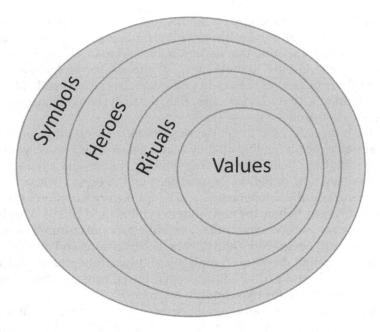

FIGURE 7.1
Hofstede's Four Layers of Culture

   3. Subculture

   4. Subjective Culture

In 2008, Katherine Mearns and Stephen Yule conducted a research project at the University of Aberdeen[62] where they examined behaviours among a multicultural workforce of an organisation operating in six countries, with distinctive national cultures. Utilising Geert Hofstede's cultural dimension model, they looked for a relationship between the established cultural dimensions of each geographical area, worker perceptions of the organisation's "safety culture" and at-risk behaviours.

   Prior to this study, it had been theorised by many researchers that the influence culture has on an individual varies depending on the nature of the behaviour under investigation. In other words, an individual's behaviour is influenced more by certain types of culture and less by others, depending on the situation. An example of this could be the differences between an individual's behaviour at work, at home, in a place of worship or on vacation, same person from the same national culture in four very different environments. So, the question remains . . . What layer and/or cultural type dictates a "safety culture"?

   Figure 7.2 demonstrates the belief that cultural types interact and that individual beliefs and behaviours are a product of the subjective culture where the various layers overlap. Some researchers such as Peter Dorfman and Jon

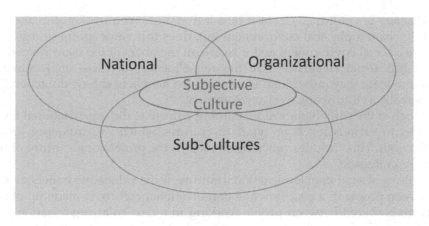

FIGURE 7.2
Cultural Types

Howell[67] believe that the dynamic nature of the individual or subjective culture will vary depending on the level of personal influence the other culture types have at any given time.

This belief was supported by the Mearns and Yule study in 2008, which concluded in part that behaviour can be considered a subjective culture which impacts other cultural types in varying degrees depending on the situation and environment. Place of Worship versus Club Med or Culture versus Climate?

Before we venture off topic and head into operational climates and work environments, let's take a minute to dissect safety as a "culture" based on the four characteristics as mentioned previously.

1. Culture is a collection of individuals who share common values, beliefs, ideas, etc.

Despite the debate, most researchers concede that, regardless of the type of culture being examined, every group differentiated from the others based on how they defined values. While values-based safety has been bantered around for the past three decades do values really depict culture and if so what type of culture?

In his book Values Based Safety Process – Improving Your Culture with Behavior-Based Safety[34], Terry McSween explores culture as a "question of balance".

McSween highlights the importance the work environment plays and the interaction between it and behaviour.

Unsafe work behavior is accordingly the result of (1) the physical environment, (2) The social environment, and (3) worker's experience with these.

If safety is a culture, is it static or is it dynamic? As a work group, we certainly share a physical environment but, does this same group necessarily share a social environment and if so are all members of the team welcomed into this social construct? Are the consequences of these environments shared or are they individual? Do some team members choose not to join? Is this really a culture or a work environment?

Much of the early literature, examining culture at the organisational level, points to various levels or types of cultures which can influence group behaviour. This includes subcultures such as the professional culture of the safety profession.

Culture at a national level, national culture, is the values and beliefs shared between people in a society while organisational culture is made up of the practices shared between people working in a particular organisation, setting or profession.

2. Culture is not individual, but individuals learn the culture of a group when they become a member of the group.

Researchers such as Killmann[68] and Dorfman[67] et al. believe that workers may be socialised into an organisational culture, which will influence decision-making and risk tolerance.

In my early research at the University of Greenwich[69], I found no link between national culture and at-risk behaviour and no measurable relationship between organisational culture and risk. So as an individual who joins a work team or organisation at what point do I become a member of this safety culture and how long does it take for me to learn the traditions, values, and rituals? More importantly, who oversees teaching me the group culture?

In most cases, I am given a safety orientation, shown a book or two on processes, policy and procedures, and perhaps even shown a video from the boss telling me how much they value my safety. Then, I enter the safety management cycle and make choices based on my own moral assessment, which has nothing to do with the group "culture".

Am I now part of this culture we call safety? Will my actions and decisions be for now and all time driven by this new mob mentality I've been introduced to? If I get transferred to another location within the organisation will the culture of the organisation come with me, or will I have to learn another culture? Likewise, when my leader quits or moves on . . . does my new leader know our language, traditions, rituals, and values?

The answer to all these questions is no. Why? Because these things are subjective at best and dependent entirely on individual circumstance, beliefs and perceptions.

3. Historical dimension is a related aspect of culture.

Safety, or a lack of it, is not a culture but, as mentioned above, the comfort zones of blaming systemic failures on workers by way of a catchphrase and the result of a professional subculture that keeps us stuck in the activity trap of managing risk with rules focused on the desired outcome and managing workers with process-driven values statements that cannot be demonstrated. All the while setting targets we know cannot be hit, all carried over from the history passed down within the professional subculture, many of them stemming back over 100 years. Not "safety culture" but "safety management subculture". Work teams do not belong to this subculture and most often shun it because it does nothing but hinders their needs to get work done on time and budget.

The term safety culture is nothing more than a passive-aggressive means of hanging on to our history to justify doing the same thing and not achieving different results. It is a soft way of continuing to blame the worker.

4. Culture has four distinct layers.

No matter which definition of safety we adhere to, the noun or the verb, we come up about three layers short of a culture.

As Terry McSween demonstrated in Figure 7.3, culture is a balancing act between stated processes and demonstrated values. These processes and values are set by leadership, usually with the guidance of a safety professional, to control worker behaviour and have nothing to do with a workers' personal values, rituals, heroes or symbols.

The only thing workers bring to the table is their personal consequence history dealing with the historic actions of the professional subculture of the safety profession which most often creates moral fatigue as demonstrated in the safety management cycle discussion in Chapter 6.

The challenge is not a lack of safety culture or a bad safety culture. The challenge lies in the systemic bad habits inherited, taught, and coveted within a professional subculture called safety management.

FIGURE 7.3
Terry McSween's Standard Elements of Organizational Culture

Safety Profession Subculture

According to the Oxford English Dictionary, subculture means "an identifiable subgroup within a society or group of people, especially one characterised by beliefs or interests at variance with those of a larger group". The subculture differentiates itself from the larger group or parent culture to which it belongs by developing its own norms, values, rituals and heroes while maintaining some of the founding principles of the parent group.

In reference to the model shown in Figure 7.2, while a professional group will share a common set of values the individual behaviours of the members are impacted by the interaction between the professional culture, national culture and organisational culture. It is the overlap or interaction, where the culture types align, that dictates behaviour.

This would explain the divisions we often see between those who subscribe to different ideologies regarding safety management: the rules-based practitioners versus the soft skills practitioners. Those who come from industry versus those operating in academia versus those deciding and setting policy. Each group's behaviour is impacted by the other types of cultures they belong to, that is, national, organisational and perhaps industrial or institutional. The interaction between these various cultural types is a big part of what drives the division between groups yet, despite this, as a subculture on its own the professional culture of safety practitioners continues to share history and to some extent values, rituals and symbols. This could be what drives our bad habits and creates the comfort zones we refuse to leave behind.

Another consideration at play is the ethical environment they operate in. Without diving too deep into ethical climates, it is important here because it impacts both the methodologies we subscribe to and the context we implement them in. It can also impact the success or failure of programmes in relation to the climate of the subjective culture created by the interaction.

For the purpose of this discussion, we will group ethical climates into three:

1. Egoist
2. Benevolent
3. Principled

An egoist climate is one where decisions are formulated based on "what's good for me" or "what's good for my team". In a benevolent climate, we consider "what is best for the common good", and in principled climates, our decision-making requires rules and order.

While this is a very simplistic explanation of ethical climates, it serves a purpose here for two reasons: First, it demonstrates why our comfort zones fail to create sustainable improvements in safety performance, and second, why new concepts get applied out of context.

Our comfort zones, for the most part, are principles based. Rules-based safety focuses on desired outcomes aiming for compliance. Competing with this is the holistic era of safety we are now in are the soft skills and conceptual models aimed at discretionary effort and interdependency. The combination of these two is then most often introduced to people working in egoist climates. This is where incentives are effective, which can account for shortcuts and impact behaviours around things such as reporting and compliance.

As a professional subculture, the safety profession is in a constant state of conflict. Our comfort zones and rituals are very principled in nature, and we cling to symbols created by heroes from another time, place and subcultures (professions), and we utilise this common history in egoist environments framed in benevolent messaging.

Understanding and identifying the ethical climate we are operating in defines operational realities and is a crucial step in fact finding and developing corrective actions based on fact.

Governments and industrial organisations continue to spend millions searching for the holy grail they call safety culture. Trying to unearth the ever-elusive means to measure and quantify forgetting that not everything that can be counted, counts and not everything that counts can be counted. This search for perfection is always worker focused based on desired results. During the writing of this book, I have seen no less than three press releases from various health and safety associations proudly announcing yet more money allotted to measure something that does not exist.

What I have never seen, and what keeps us utilising bad habits from our comfort zones, is money allotted to examine the professional subculture of safety. What we fail to do is shine the spotlight on ourselves. Our comfort zones and bad habits are so deeply rooted that, regardless of the subjective culture we split into, we all carry forward those rituals, heroes and symbols from the parent culture of safety management.

We continue to apply tools out of context. We continue to dress principles up as values and frame blame the worker as benevolence all the while keeping a watchful eye on "what's in it for me?"

A Last Word on Professional Values and Applied Ethics

Values are the stuff that applied ethics is made of, and therefore stepping out of our comfort zones requires an applied ethics approach in examining the professional subculture of the occupational health and safety profession.

Applied ethics in safety is not about what is right and what is wrong. It is not about that policy or procedure you copy–cut–paste to call your own. It is not about claiming to be something you are not or selling skills or qualifications you do not possess. These things are better left to the professional associations.

Utilising an applied ethics lens to examine the profession is not even about codes or oaths. It is about turning the lens on the profession and examining what is driving our decisions both in practice and in regulation. It provides a means to facilitate conversations outside the comfort zone and work towards breaking bad habits. It creates a safe space in which to identify blind spots created by the activity trap and hold difficult conversations. It allows us to reflect upon the discomfort of breaking out of the comfort zone. It also provides a framework to begin diagnosing the ethical climates created by our profession.

Like the search for a tangible measurable safety culture, there is a great deal of resources being poured into standardised safety management systems, creating international reciprocity between national professional qualifications and defining what a safety professional is. While this is important work, there is a great deal of time and money being spent on splinter groups and empire builders to break away from others and define themselves as different and perhaps somewhat better than the old guard, yet we bring with us the parent groups values, rituals, heroes and symbols.

We fail to leave the bad habits behind. Why? Because we, as a profession, continue to focus on desired results utilising the seven bad habits to address our lack of success.

Over the past 30-plus years, I have attended and presented at national and international conferences. The themes always change but the focus remains the same: "How can we improve the safety we do to people" or more specifically, "How can we stop the behaviours they do to us?"

I have heard many great ideas and have shared many experiences, but what I have not seen is a change in the way we govern, regulate and qualify safety success.

Perhaps it is time we step back from trying to improve the work environments of people who know how to do the work and shine a light on the occupational health and safety profession. Maybe it is time we stopped asking why workers are making the bad choices they sometimes make and examine why we make them choose.

Safety culture has run its course but as long as there is research money available to find it and regulators continue to buy it as a human factor explaining failure, we will continue to create moral fatigue among workers driving them to moral disengagement.

Stepping outside the comfort zone and leaving the bad habits behind will not be easy, but we will never get there if we collectively do not make the first step. It is time to stop looking at worker behaviour and start questioning ours. As a profession, we need to examine our subculture and how it interacts within various environments. We need to identify the root of the research–regulation–implementation gap and cut it out of our practice. We need to focus on uncomfortable truths based on information derived from our unconsciously demonstrated values and be the change we expect to see on the job.

First and foremost, we need to start the difficult conversations and step out of the shadows of our comfort zones.

References

1   Shewhart WA (1980 [1931]) *Economic Control of Quality of Manufactured Product*. Milwaukee: American Society for Quality, ISBN: 978-0873890762. OCLC 7543940. 50th Anniversary Commemorative Reissue. Originally published 1931. New York: Van Nostrand.

2   Moen RD, Norman CL (2010) *Circling Back – Clearing Up Myths About the Deming Cycle and Seeing How It Keeps Evolving*, accessed online at: https://deming.org/wp-content/uploads/2020/06/circling-back.pdf, accessed 21 December 2022.

3   Moen Ronald D (1991) *Improving Quality Through Planned Experimentation*. New York: McGraw-Hill College, ISBN: 0070426732.

4   Agnew J, Daniels A (2010) *Safe by Accident, Take the Luck Out of Safety, Leadership Practices That Build a Sustainable Safety Culture*. Atlanta: Performance Management Publications, Aubrey Daniels International Inc., ISBN: 0937100188.

5   Ishikawa K (1985) *What Is Total Quality Control?* (Lu DJ trans.). Englewood Cliffs, NJ: Prentice-Hall Inc.

6   McAleenan C (2016) *Operation Analysis and Control: A Paradigm Shift in Construction Safety Management – Volume 1 – Critical Review*. Pontypridd, Wales: University of New South Wales.

7   Ritchie M (2015) *Strategic Step Change in Operational Safety Climates*, accessed online at: www.tri-lenssafety.com.

8   Busch C (2016) *Safety Myth 101 – Musings on Myths, Misunderstandings and More*. First Edition. Adapted for Kindle, accessed online at: www.mindtherisk.com, ISBN: 978-82-690377-1-5.

9   Bird FE, Germain GL, Clark MD (2014) *Loss Control Management – Practical Loss Control Leadership*. Third Edition. Katy, TX: DNV GL, ISBN: 0-88061-054-9.

10 Accessed online at: www.history.com/topics/roaring-twenties/prohibition, accessed 5 February 2023.

11 Heinrich HW (1941) *Industrial Accident Prevention: A Scientific Approach*. Second Edition. First published 1931. New York and London: McGraw-Hill, accessed online at: https://archive.org/details/dli.ernet.14601/page/n5/mode/2up, accessed 6 February 2023.

12 Bahn ST (2014) *OHS Management: Contemporary Issues in Australia*. Prahran: Tilde University Press.

13 Taylor FW (2005 [1911]) *The Principals of Scientific Management*. eBook First Edition. Fairfield, IA: 1st World Library – Literary Society, ISBN: 1-4218-0540-5.

14 Health and Safety Executive (HSE) (2006) *Successful Health and Safety Management*. First published 1991. London: HMSO, ISBN: 0-7176-1276-7.

15 Minter S (1998, July [2016]) *The Birth of OSHA*. Occupational Hazards, 59. National Safety Council, Accident Facts; Chicago: National Safety Council.

16 ARPANSA (2022) *History of Safety*. Australian Government © Commonwealth of Australia, accessed online at: www.arpansa.gov.au/, accessed 22 February 2023.

17 ASQ (2022) *History of Total Quality Management*. American Society for Quality, accessed online at: https://asq.org/quality-resources/total-quality-management/tqm-history#:~:text=W.,considered%20the%20origin%20of%20TQM, accessed 11 February 2023.

18 Kontogiannis T, Leva MC, Balfe N (2016–2017, December) Total Safety Management: Principles, Processes and Methods. *Safety Science*, 100(Part B); 128–142. Elsevier, accessed 10 February 2023.

19 mgnep (2019) *Management & Protection Systems – Documents: BS 8800*, accessed online at: www.mgnep.com, accessed 9 February 2023.

20 Gasiorowski-Denis E (2016) *ISO 45001 on Occupational Health and Safety Has Been Approved for Draft International Standard Public Consultation*, accessed online at: www.iso.org/2016/02/Ref2012.html, accessed 10 February 2023.

21 Health and Safety Executive (HSE) (2006) *Successful Health and Safety Management*, page 14 Diagram 3. First published 1991. London: HMSO, ISBN 0-7176-1276-7.

22 Duteil B (2017) *OHSAS 18001 – What You Need to Know*. Ecratum, accessed online at: https://blog.ecratum.com/ohsas-18001-what-you-need-to-know#:~:text=OHSAS%2018001%20fundamentals&text=The%20OHSAS%2018001%20is%20based,wheel%20or%20the%20Shewhart%20cycle, accessed 10 February 2023.

23 ISO (2022) *ISO 45001:2018 – Occupational Health and Safety Management Systems – Requirements With Guidance for Use*. International Organisation for Standards, accessed online at: www.iso.org/standard/63787.html#:~:text=ISO%2045001%3A2018%20specifies%20requirements,proactively%20improving%20its%20OH%26S%20performance, accessed 14 February 2023.

24 Tarlengco J (2022) *Safety Management Systems (SMS)*. © SafetyCulture, accessed online at: https://safetyculture.com/topics/safety-management-system, accessed 13 February 2023.

25 Nolan J (2019) *ISO 45001 Requirements and Structure*, accessed online at: https://advisera.com/45001academy/blog/2019/02/05/iso-45001-requirements-and-structure/, accessed 10 February 2023.

26 St John Holt A (2003) *Principals of Health and Safety at Work*. Third International Edition. Southampton: Hascom Network Limited, ISBN: 0-9013-5733-2.

27 Taylor FW (2005 [1911]) *The Principals of Scientific Management*. eBook First Edition. Fairfield, IA: 1st World Library – Literary Society, ISBN: 1-4218-0540-5.

28 Nickerson C (2022) Bureaucratic Management Theory of Max Weber. *Simply Sociology*, accessed online at: https://simplysociology.com/bureaucratic-theory-weber.html, accessed 14 February 2023.

29 St John Holt A (2003) *Principals of Health and Safety at Work*. Third International Edition, page 2. Southampton: Hascom Network Limited, ISBN: 0-9013-5733-2.

30 Rhodes A (2015) *A Brief Summary of the Long History of Risk Management*, accessed online by Ventiv Technology at: www.ventivtech.com/blog/a-brief-summary-of-the-long-history-of-risk-management, accessed 14 February 2023.

31 Hurst NW (1998) *Risk Assessment – The Human Dimension*. Cambridge: The Royal Society of Chemistry Information Services, ISBN: 0-8540-554-6.

32 Health and Safety Executive (HSE) (2006) *Successful Health and Safety Management*, page 93. First published 1991. London: HMSO, ISBN 0-7176-1276-7.

33 CCOHS (1997–2023) *OSH Answers Fact Sheets – Hazard and Risk – Risk Assessment*. Canadian Centre for Occupational Health & Safety, accessed online at: www.ccohs.ca/oshanswers, accessed 12 January 2023.

34 McSween T (2003) *Values-Based Safety Process – Improving Your Culture With Behavior-Based Safety*. Second Edition. Hoboken, NJ: John Wiley & Sons Inc., ISBN: 0-471-22049-3.

35 Hurst NW (1998) *Risk Assessment – The Human Dimension*, page 66. Cambridge: The Royal Society of Chemistry Information Services, ISBN: 0-8540-554-6.

36 St John Holt A (2003) *Principals of Health and Safety at Work*. Third International Edition, page 7. Southampton, England: Hascom Network Limited, ISBN 0-9013-5733-2.

37 Howell GA, Ballard G, Abdelhamid TS, Mitropoulos P (2002) *Working Near the Edge: A New Approach to Construction Safety*. Submitted for Inclusion in the Proceedings of the 10th Annual Conference of the International Group for Lean Construction, Proceedings IGLC-10, Gramado, Brazil.

38 Funtowicz SO (2021) *A Quick Guide to Post-Normal Science*. Integration and implementation Insights, accessed online at: https://i2insights.org/2021/10/19/guide-to-post-normal-science, accessed 28 December 2022.

39 Buschke FT, Botts EA, Sinclair SP (2019) Post-Normal Conservation Science Fills the Space Between Research, Policy, and Implementation. *Conservation Science and Practice*, 1; e73, https://doi.org/10.1111/csp2.73.

40 Health and Safety Executive (HSE) (2006) *Successful Health and Safety Management*, page 93. First published 1991. London: HMSO, ISBN 0-7176-1276-7.

41 Quilley A (2020) *Lies Statistics Can Tell You*, accessed online at: www.linkedin.com/pulse/lies-statistics-can-tell-you-alan-quilley-crsp?trk=public_profile_article_view.

42 Public Citizen (2011) *As Transocean Touts Its Best Year in Safety, Government Has Amnesia About Deepwater Horizon Explosion and Oil Spill*. Public Citizen Foundation, accessed online at: www.citizen.org/news/as-transocean-touts-its-best-year-in-safety-government-has-amnesia-about-deepwater-horizon-explosion-and-oil-spill/, accessed 7 January 2023.

43 Eastman C (1910) *Work-Accidents and the Law – Volume 2 – The Pittsburgh Survey*. New York: Russell Sage Foundation Publications. ISBN None, accessed online at: www.russellsage.org/sites/default/files/Eastman%26Kellog_Work%20Accidents_0.pdf.

44 Edwards v National Coal Board (1949) – Reasonably Practicable – Definition, the Quantum of Risk Test accessed online at: http://www.safetyphoto.co.uk/subsite/case%20e%20f%20g%20h/edwards_v_national_coal_board.htm, accessed 12 January 2023.

45 Heinrich HW (1931) *Industrial Accident Prevention: A Scientific Approach*. New York: McGraw-Hill. OCLC 571338960.

46 Bird FE (1969) *Loss Control Management: Practical Loss Control Leadership*. Revised Edition 1996. Loganville: Det Norske Veritas, ISBN: 0880610018.

47 Bird FE, Germain GL, Clark MD (2014) *Loss Control Management – Practical Loss Control Leadership*. Third Edition, page 6. Loganville: DNV GL, ISBN: 0-88061-054-9

48 HSE (1993) *The Cost of Accidents at Work*. Second Edition. London: Health and Safety Executive Paperback, ISBN: 0-7176-1343-7.

49 Health and Safety Executive (HSE) (2006) *Successful Health and Safety Management*, page 56. First published 1991. London: HMSO, ISBN 0-7176-1276-7.

50 Jakulin A (2007) *Quotee of the Day, Sir Josiah Stamp*, accessed online at: https://statmodeling.stat.columbia.edu/2007/03/13/quotee-day-sir-josiah-stamp/, accessed 28 December 2022.

51 OSHA (2019) *Using Leading Indicators to Improve Safety and Health Outcomes*. Occupational Safety and Health Administration – US Department of Labor, accessed online at: www.osha.gov/leadingindicators, accessed 3 January 2023.

52 Smith S (2012) Caterpillar: Using Proactive Leading Indicators to Create World-Class Safety. *EHS Today*, accessed online at: www.ehstoday.com/safety/caterpillar-using-proactive-leading-indicators-create-world-class-safety, accessed May 2019.

53 Michaels D (2016, March 17) *Year One of OSHA's Severe Injury Reporting Program: An Impact Evaluation*, page 7. OSHA, accessed online at: www.osha.gov/sites/default/files/severe-injury-2015.pdf, accessed 19 November 2022.

54 Krause TR (2018) *A New Paradigm for Fatality Prevention*. DEKRA North America, Inc., accessed online at: www.dekra-insight.com.

55 USW (1998) *The Steelworker Perspective on Behavioral Safety- Comprehensive Health and Safety vs. Behavior-Based Safety*. Pittsburgh, PA: United Steelworkers of America Health, Safety & Environment Department, accessed online at: assets. usw.org/resources/hse/Resources/uswbbs.pdf, accessed 5 September 2013.

56 Ritchie M, Thachuk A (2013) *The Application of Safety Ethics in Operationalizing Behavior Based Safety*. Tri-lens Safety, accessed online at: www.trilenssafety.com.

57 Cooper D (2001) *Improving Safety Culture – a Practical Guide*. Second Edition, page 216. Publisher: Applied Behavioural Sciences, ISBN: 0-471-95821-2.

58 Conklin T (2020) *Learning Teams from Dr Todd Conklin*. Learning Teams Inc., accessed online at: www.learningteamscommunity.com/learning-teams-101.

59 Sutton B, McCarthy G, Robinson B (2020) *The Practice of Learning Teams*. Santa Fe: Learning Teams Inc., ISBN: 9798665374321.

60 Hofstede G (2001) *Cultures Consequence: Comparing Values, Behaviours, Institutions and Organisations Across Nations*. Second Edition. Thousand Oaks, CA: Sage Publications.

61 Arnett J (2002) The Psychology of Globalisation. *American Psychologist*, 57(10); 774–783. American Psychological Association, Inc., https://doi.org/10.1037//0003-066X.57.10.774.

62 Mearns K, Yule S (2008) The Role of National Culture in Determining Safety Performance: Challenges for the Global Oil and Gas Industry. *Safety Science*, https://doi.org/10.1016/j.ssci.2008.01.009.

63 Herskovitz M (1972) *Cultural Relativism: Perspectives in Cultural Pluralism*. New York: Random House.

64 Shweder R, LeVine R (1984) *Culture Theory: Essays on Mind, Self and Emotion*. New York: Cambridge University Press.

65 Hofstede G (1991) *Cultures and Organizations: Software of the Mind*. New York: McGraw-Hill, as cited by Rene Olie (1995).

66 Leidner D, Kayworth T (2006) A Review of Culture in Information Technology; Toward a Theory of Information Technology Culture Conflict. *MIS Quarterly*, 30(2); 357–399.

67 Dorfman W, Howell J (1988) Dimensions of National Culture and Effective Leadership Pauern, Hofstede Revisited. *Advancement in Comparative Management*, 3; 127–150.

68 Killmann R, Saxton M, Serpa R (1986) Issues in Understanding and Changing Culture. *California Management Review*, 29(2); 86–96.

69 Ritchie M (2009) *Examining the Relationship Between National Cultures, Personality Traits, Perceptions and Behaviours*. Avery Hill Campus, Bexley Road, Eltham and London: University of Greenwich.

Index

Printed in the United States
by Baker & Taylor Publisher Services